Lecture Notes in P

T0238130

The Lecture Notes in Physics

The series Lecture Notes in Physics (LNP), founded in 1969, reports new developments in physics research and teaching – quickly and informally, but with a high quality and the explicit aim to summarize and communicate current knowledge in an accessible way. Books published in this series are conceived as bridging material between advanced graduate textbooks and the forefront of research and to serve three purposes:

- to be a compact and modern up-to-date source of reference on a well-defined topic

- to serve as an accessible introduction to the field to postgraduate students and nonspecialist researchers from related areas

- to be a source of advanced teaching material for specialized seminars, courses and schools

Both monographs and multi-author volumes will be considered for publication. Edited volumes should, however, consist of a very limited number of contributions only. Proceedings will not be considered for LNP.

Volumes published in LNP are disseminated both in print and in electronic formats, the electronic archive being available at springerlink.com. The series content is indexed, abstracted and referenced by many abstracting and information services, bibliographic networks, subscription agencies, library networks, and consortia.

Proposals should be sent to a member of the Editorial Board, or directly to the managing editor at Springer:

Christian Caron
Springer Heidelberg
Physics Editorial Department I
Tiergartenstrasse 17
69121 Heidelberg / Germany
christian.caron@springer.com

J. Ferreira
C. Dougados
E. Whelan (Eds.)

Jets from Young Stars

Models and Constraints

 Springer

Editors

Jonathan Ferreira
Catherine Dougados
Université Joseph Fourier
Laboratoire d'Astrophysique de Grenoble
BP 53
38041 Grenoble, France
Jonathan.Ferreira@obs.ujf-grenoble.fr
Catherine.Dougados@obs.ujf-grenoble.fr

Emma Whelan
Dublin Institute for Advanced Studies
5 Merrion Square
Dublin 2, Ireland
ewhelan@cp.dias.ie

J. Ferreira, C. Dougados and E. Whelan (Eds.), *Jets from Young Stars*, Lect. Notes Phys.
723 (Springer, Berlin Heidelberg 2007), DOI 10.1007/978-3-540-68035-2

Marie Curie Research Training Networks Contract, JETSET
(Contract No MRTN-CT-2004-00559)

ISSN 0075-8450
ISBN 978-3-642-08769-1
e-ISBN 978-3-540-68035-2

Springer is a part of Springer Science+Business Media
springer.com
© Springer-Verlag Berlin Heidelberg 2007
Softcover reprint of the hardcover 1st edition 2007

A
Cover design: eStudio Calamar S.L., F. Steinen-Broo, Pau/Girona, Spain

Preface

Crucial to the development of a protostar and active for most of the pre-main sequence phase, the mass outflow phenomenon is the most spectacular manifestation of star formation. The generation of jets from young stars involves a complex interplay, still poorly understood, between gravity, turbulence, and magnetic forces that may have important implications on the removal of excess angular momentum from the star-disk system as well as on conditions for planet formation. These jets also create shocks and ionization fronts that propagate into the surrounding medium, thereby providing feedback that can affect both the cloud structure and chemistry, how the cloud evolves, and hence the future generation of young stars. An ultimate understanding of jets from young stars, their generation, and interaction with their parent cloud is a vital part of any unified theory of Star Formation.

The motivation of the four-year JETSET (Jet Simulations Experiments and Theory) Marie Curie Research Training Network is to build an interdisciplinary European research and training community focused on the study of jets from young stars, at the confluence of astrophysical observations, theoretical and computational modelling, laboratory experiments and Grid technology. The network scientific goals will focus on understanding (i) the driving mechanisms of jets around young stars; (ii) the cooling–heating processes, instabilities and shock structures in stellar and laboratory jets; and (iii) the impact of jets on energy balance and star formation in the galactic medium. Important to these overall goals are the series of JETSET schools dedicated to the training of our young researchers in key jet topics.

This book is a collection of the lectures from our first school, Jets From Young Stars: Models and Constraints, held in Villard de Lans France in January 2006. Central to our understanding of jets is the identification of the launching and collimation mechanisms. The strong correlation of accretion and ejection signatures together with the high ejection efficiencies and small collimation scales observed have led to the identification of magneto-centrifugal processes as the main jet driving agent. The aim of the first JETSET school is to provide a solid background in magneto-hydrodynamic

(MHD) steady jet models as well as a good knowledge of the overall jet observational properties supporting them.

The schools to follow will tackle the following key issues. The second JET-SET school, hosted by the Florence node, will review modern high-angular resolution observational techniques, which have allowed us to resolve the jet launching region (within 100 AU of the protostar). The third JETSET school (Torino) will be devoted to numerical simulation techniques, a key tool to study the complex interaction of the jet with its surroundings as well as jet instabilities. The 4th JETSET school (Porto) will deal with plasma diagnostics and heating/cooling processes, crucial ingredients to link constraints brought by observations/experiments to model predictions/simulations. The fifth and last JETSET school (Dublin) will introduce students to-state-of-the-art computing grid technology and its application to large-scale jet simulations. In particular the results of the numerical simulations of jets, which are beyond the scope of this first book, will be discussed at the Dublin school.

The book is structured as follows. The first half is devoted to general observational constraints, starting with a presentation of the outflow phenomenon in young stars by Tom Ray. Sylvie Cabrit then reviews arguments that have led us to identity magneto-centrifugal processes as the main jet driving mechanism. Finally, our current understanding of the star-disc magnetic interaction, critical to both the accretion and outflow processes, is presented by Silvia Alencar. The second half of the book reviews theoretical knowledge of MHD processes pertinent to the jet launching mechanism in young stars. Guy Pelletier provides a general introduction to magneto-hydrodynamics, including its validity conditions and a brief overview of the reconnection process. Caroline Terquem's contribution details the physics of Standard Accretion Discs. Although such disc models do not produce jets, they provide a description of the observed accretion phenomenon, which relies also on MHD processes. Then, Kanaris Tsinganos introduces the physics of steady-state MHD outflows, from the basic concepts and equations to modern self-similar solutions. The next three lectures detail the various classes of steady magnetic wind models currently discussed in the context of protostellar jets: Thibaut Lery introduces the transit flow model, aimed at explaining the large molecular ouflows observed in the early stages of star formation and briefly discusses asymptotic jet equilibria. Jonathan Ferreira's lecture covers the physics of Jet Emitting Discs and their related jets, usually referred to in the literature as disc-winds or X-winds. Finally, Christophe Sauty introduces the main properties of stellar winds.

The editors would like to thank all the lecturers for their excellent presentations and contributions to this book. We are also thankful to all school participants who brought a studious but very enjoyable atmosphere to the week in Villard. Last but not least, we would like to acknowledge the huge amount of work done by Sandrine Vignon, Richard Mourey, and Claudio Zanni, who helped us in the organization of this school. Many thanks also to Eileen Flood who, together with Sandrine, cheerfully managed all practical aspects during the week, including getting the sky shoe size from 80 eager participants at coffee break ...

Dublin Institute for Advanced Studies   *Emma Whelan*
Laboratoire d'Astrophysique de Grenoble   *Catherine Dougados*
June, 2006   *Jonathan Ferreira*

Contents

Part I

Observational Constraints

The First Three Million Years

Tom Ray

School of Cosmic Physics, Dublin Institute for Advanced Studies, 5 Merrion Square, Dublin 2, Ireland
tr@cp.dias.ie

1 In the Beginning

In Mark Twain's *The Adventures of Huckleberry Finn* there is a passage where Huck is describing his trip down the Mississippi by raft with Jim and he says "We had the sky up there, all speckled with stars, and we used to lay on our backs and look up at them, and discuss about whether they was made or only just happened". Well of course we know they are made but how? That is a question that has endured for many hundreds if not thousands of years. Modern ideas on the subject could be said to have begun with two gentlemen: Immanuel Kant and Pierre Simon Laplace. Immanuel Kant (Fig. 1) was born

T. Ray: *The First Three Million Years*, Lect. Notes Phys. **723**, 3–19 (2007)
DOI 10.1007/978-3-540-68035-2_1

Fig. 1. Immanuel Kant was the first to propose the nebular hypothesis for the formation of the Sun and planets, a theory that was later developed mathematically by Laplace. It would be over two hundred years before his ideas could be verified

in 1724 in modern day Kaliningrad, Russia. At that time however the city was within East Prussia and known as Königsberg. Although we remember Kant today as one of the founders of modern philosophy, he also had a passionate interest in Nature, and Astronomy in particular. He read for example Newton's *Principia* and studied the effects of the tides on the Earth's rotation. In 1755, Kant published his famous book *Allgemeine Naturgeschichte und Theorie des Himmels* (Universal Natural History and Theory of the Heavens) in which he expounded the nebular hypothesis. He proposed that the Solar System formed from a collapsing cloud of gas. As it collapsed, he suggested, it would spin-up and flatten into a disk. The central region formed the Sun, and the left-over material in the disk the planets. Kant was not a mathematician, and it was some years later before Pierre Simon Laplace, born in 1749 in Normandy, gave the theory a mathematical basis in his *Exposition de système du monde* in 1796. There was however a fundamental difference between Laplace's and Kant's models: Laplace thought that the planets formed from the centrifugal ejection of rings of excess matter as the Sun condensed. In contrast Kant suggested the planets formed directly out of the swirling gas. Modern ideas, as we shall see, are thus more in line with Kant than Laplace.

Much of the early part of the twentieth century was devoted to understanding how stars actually work (the great battles, for example of Eddington and Jeans) rather than how they form. In the nineteenth century however

Lord Kelvin (then William Thomson) and Hermann von Helmholtz estimated, using the then modern Theory of Thermodynamics, how long the Sun would shine using its thermal energy reservoir alone. The result is well known, i.e. a few million years, a period known as the Kelvin-Helmholtz timescale. This timescale contrasted sharply with the age of the Earth as estimated from geological studies. In due course this disparity would lead to the idea that the Sun, and other stars, shone through nuclear fusion rather than through gravitational contraction. The concept of the Kelvin-Helmholz timescale is of course still useful, at least, to star formation studies, since it corresponds approximately, and for obvious reasons, to the time it takes a star to form from a collapsing core within a molecular cloud.

The discovery of molecular clouds can, in a sense, be traced back to the 18th century. In the Spring of 1784 William Herschel was counting the stars in various parts of the sky using his giant telescope. His sister Caroline patiently took notes by his side. At one point during the night, William turned to Caroline and said "Hier ist wahrhaftig ein Loch im Himmel!" (Here is certainly a hole in the Heavens!) The dark nebulae, or molecular clouds as we prefer to call them had been discovered. Herschel did not attach much significance to his finding, believing he had discovered nothing more than an actual gap in the stars. In contrast Caroline thought more of it and went on, with the assistance of her nephew John Herschel, to compile the first catalogue of dark nebulae many years later.

The realisation that dark nebulae in the Milky Way, such as the Coalsack, were not actual holes in the distribution of stars, but instead regions blocking our view of the stars beyond, did not come until the early part of the twentieth century. Studies by astronomers, such as Edward Barnard at Lick Observatory [9] and Max Wolf [52] in Heidelberg, showed them to be true astronomical objects. Later in the twentieth century, Bart Bok found a number of dark globules, which now bear his name, near HII regions. He suggested, in a paper published in Astrophysical Journal in 1947 with E.F. Reilly [11], that these globules were in the process of collapsing to form stars.

Around the same time Alfred Joy [21] discovered a whole new class of irregular variable stars which we now refer to as T Tauri stars. Despite their association with the dark nebulae, Joy did not realise their significance to star formation. Instead this was left to Viktor Ambartsumian in 1947 [1]. In the 1950s and 1960s, Chushiro Hayashi developed models for the evolution of pre-main sequence stars in the Hertzprung Russell diagram. He was the first to realize that low mass stars follow almost exclusively convective tracks onto the main sequence [26]. Along such a track, the surface temperature remains almost constant as the radius, and hence the luminosity, decreases.

In closing this historical introduction, it is worth reiterating that the basic model of how stars and planets form has been with us for many years. The difficulty however has been in testing it. Even up until fairly recent times, it was seriously suggested that the planets in the Solar System arose when the Sun had a strong gravitational encounter with another star. Of course this theory can be easily dismissed: such encounters are very rare indeed in the Milky Way and besides the Sun and the planets are approximately the same age. To check Kant's and Laplace's hypothesis however requires imaging solar systems in creation. The problem in doing this is twofold: the nearest star formation systems are so far away that even a nebula that extends out to the orbit of Pluto is only just about observable with standard ground-based equipment. Even more important is the obscuring effects of dust since it makes observations of the earliest stages of the star formation process impossible, at least in the optical regime. Dust, on the other hand is fundamental to the star formation process: without it, as pointed out below, we would not have molecular clouds. Note however that dust only constitutes 1% by mass of a molecular cloud.

Finally I should remark on the title. A number of years ago, Stephen Weinberg wrote his famous book *The First Three Minutes* in which he described what happened in the early phases of the Big Bang. In contrast, around the same period, we knew virtually nothing about the *first three million years* in the life of the Solar System. The picture has changed dramatically in the past few years largely due to the availability of modern instrumentation, particularly at infrared and mm wavelengths. With these ideas in mind and an apology to Weinberg, we begin our brief overview of the star formation process. Much of the detail will be filled-in by subsequent chapters not only in this book but in the ones that follow it in the JETSET School Series. We start by saying a few words about the sites of star formation, i.e. molecular clouds.

2 Molecular Clouds

With the development of mm waveband astronomy, it became clear that the Milky Way, as well as other spiral galaxies, contained vast quantities of star-making material in the form of molecular clouds. The clouds themselves could be identified with the dark nebulae noticed by Herschel many years before. Of course the lack of transparency in a dark cloud has nothing to do with the presence of molecules but, as already mentioned, is instead due to dust. Dust, however, and molecular clouds are intimately related. Without dust most star forming molecular clouds would cease to exist since the dust protects the molecules against dissociating interstellar uv radiation (uv radiation if effectively blocked once the optical extinction $A_V \geq 1$ and most clouds have A_V values measured in tens to a few hundred magnitudes). Moreover not only does dust protect molecules but dust grains are also sites were even the simplest of molecules, such as H_2 can form. Aside from H_2, more exotic species such as CO, NH_3, amino acids, ethanol, and poly-aromatic-hydrocarbons (PAHs) are found. In fact it was once calculated that the Sagittarius B2 Cloud at the center of our galaxy contains enough ethanol for 10^{28} bottles of whiskey. Proof-positive, as one wag put it, that there is a bar at the center of the Milky Way!

Molecular clouds can be divided up into two types: small molecular clouds (SMCs) such as the Taurus-Auriga or Ophiuchus clouds (see Fig. 2) and giant molecular clouds (GMCs), e.g. Orion (see Fig. 3). They differ in many properties other than the obvious ones of mass and size. SMCs typically contain a few hundred solar masses of gas and dust and sizes measured in a few tens of parsecs. They do not contain young massive OB stars and they are relatively cool ($T \approx 10$–$20\,K$). In contrast GMCs have masses around $10^{4-6}\ M_\odot$, sizes of approximately $100\,pc$, both massive and low mass young stars, as well as higher ambient temperatures ($T \approx 50$–$100\,K$). Moreover the distribution of GMCs and SMCs is different: GMCs are confined to the spiral arms whereas SMCs are found not only in the spiral arms but in the inter-arm regions as well. Our galaxy is host to some 4,000 giant molecular clouds and many more small molecular clouds.

What is obvious from even a cursory inspection of a molecular cloud is its filamentary and clumpy nature. Note that such clouds are usually "probed" using rotational transitions of molecules like CO rather than the much more

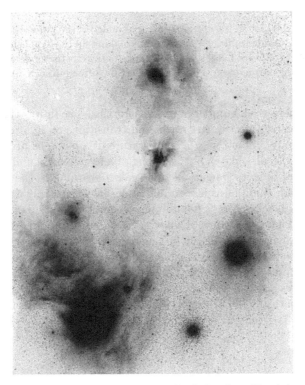

Fig. 2. Anglo Australian Telescope image of the Ophiuchus Cloud showing not only stars in formation but dark regions abundant in molecules and dust. Although the dust obscures our optical view of the star formation process, without it a molecular cloud could not form. Image courtesy of David Malin

Fig. 3. Palomar 200-inch Telescope near-infrared image of the Orion Cluster centred on the Trapezium. Aside from the massive stars making up the Trapezium, there are enormous numbers of lower mass solar-like stars. As the cluster mode of star formation seems to be the norm; it seems likely that the Sun formed in such an environment

common H_2. There are a number of reasons for this. The first is that, because of its low moment of inertia, the rotational transitions of H_2 are well above the ground state in a cloud with a temperature of a few tens of Kelvin. The second is that molecules like CO have strong dipole moments, unlike H_2 which is symmetric and thus its transitions are weak. Returning to the structure of a molecular cloud: clumpiness is seen on many scales. Not only does a typical cloud possess filaments, clumps (containing several hundred solar masses) and cores (made up of a few or tens of solar masses) but also diffuse structure.

Of course, gravity is relentlessly pulling on molecular clouds trying to force them to collapse: it is thus reasonable to ask what supports such a cloud? Cloud temperatures, as I have mentioned, are very low, and correspondingly thermal pressures.

This is dramatically illustrated by considering the Jeans mass for the cloud as a whole. M_J, is given by:

$$M_J = 1.6 \left(\frac{T}{10\,K} \right)^{3/2} \left(\frac{n}{10^4\,cm^{-3}} \right)^{-1/2} M_\odot \tag{1}$$

where T is the temperature of the cloud and n is its average number density. If we put in the typical temperature of a GMC and its density, we get a Jeans mass of around a few hundred $M_\odot$. In contrast the GMC can contain 10^6 $M_\odot$! A clue, however, as to what may be providing the support is given by observations of mm rotational transitions such as the J=1-0 CO line. The widths of such lines are found to be several km s^{-1} whereas the thermal width, at a temperature of a few tens of Kelvin, should be much less than one km s^{-1}. It is generally agreed that this line broadening is due to turbulence since, as discovered by Larson, the velocity dispersion observed is a function of the scale on which you look [37]. This is precisely what would be expected from turbulence (for example from a Kolmogorov-type spectrum) as energy cascades from larger to smaller scales. Moreover as the velocities of the turbulence are greater than the sound speed, it is supersonic and thus expected to generate shocks where some of its energy is dissipated.

Referring again to the structures seen in molecular clouds, I have remarked already how both clumps and cores are seen. The mass of clumps would suggest that ultimately these structures give rise to the young stellar clusters we observe such as the Orion Cluster centred on the Trapezium. In contrast cores are more likely to give rise to individual stars, binaries or multiple systems. It is interesting to note that in recent years an analysis has been made in several clouds of the distribution of masses in cores. Such studies show that the mass distribution follows closely that of the so-called Initial Mass Function (IMF) of stars, i.e. the number of stars versus mass relationship seen in young clusters [50]. This suggests that the IMF is determined at a very early stage in the star formation process through cloud fragmentation. There is however uncertainty at the high mass (OB-type) star end of the spectrum: such stars could have their final mass largely determined by competitive accretion at the center of the cluster (e.g. [12]).

Another aid to support in a molecular cloud may be its magnetic field. That such fields are present is clear from a whole host of observations. For example background stars, at the edges of molecular clouds, have enhanced linear polarization , which is expected from aligned dust grains in the cloud. In regions with higher extinction, the polarization of the mm emission from the grains can be used to infer the presence of fields (e.g. [32]). Finally field strengths can be measured directly through Zeeman splitting of various radio emission lines (e.g. due to the OH radical). Field strengths vary enormously depending on the density of the region one is observing and can range from a few μG to a few mG [13].

If the magnetic flux is sufficient to prevent collapse, the cloud is said to be sub-critical and critical if the flux is insufficient. Most cores are observed to be borderline. There are however many uncertainties in deriving the field strength, due, for example, to projection effects. Moreover, even if a cloud is sub-critical to start with, this state of affairs may not continue indefinitely. The reason is that the field is tied only indirectly to the ions in the cloud. Now the degree of ionization in the cloud is very low (perhaps 10^{-6} and primarily maintained by the penetration of cosmic rays in the darkest parts of the cloud). Coupling to the neutrals is achieved through collisions. Slippage between the neutrals and the ions occurs, a process known as ambipolar diffusion. Thus eventually the ions "leak-out", leaving the neutrals without support [39].

3 Collapse

Starting with a molecular core, how does the collapse process proceed? Understanding this process is not only observationally difficult (the free-fall timescales, for example, are very short astronomically-speaking and thus it is hard to catch a cloud "in-the-act") but also theoretically and computationally challenging. Note that the average density of a star like the Sun is 1 $g\,cm^{-3}$ whereas the typical density of a core is 10^{-20} $g\,cm^{-3}$; the change is 20 orders of magnitude! Temperatures rise from a few tens of Kelvin to tens of millions of Kelvin and size shrinks from a few light years to a few million kilometres.

Early simulations of core collapse assumed the core to be initially spherical and of uniform density and temperature. The assumption of uniform temperature is reasonable: since the presence of dust, CO and hydrogen as powerful coolants will ensure that any thermal energy, released through gravitational contraction, is efficiently radiated away. Effectively this ensures that the collapse occurs in a free-fall time. In contrast uniform density seems very unlikely: cores appear to be in quasi-hydrostatic equilibrium. Thus, as Shu proposed many years ago [48], a better starting point for modelling purposes is an isothermal sphere initially in gravitational equilibrium. It is easy to show that such a sphere has $\rho \propto r^{-2}$, i.e. has equal masses in shells of equal thickness and that the mass per unit radius is $\approx 6c_s^2/G$ where G is the gravitational constant and c_s is the sound speed. Collapse begins in the center and moves outward at the sound speed, i.e. the radius grows like $c_s \times t$. As the gas moves inwards at a speed of approximately $c_s/6$ due to the steep pressure gradient, the accretion, rate is $\approx c_s^3/G$ or around $10^{-6} M_\odot\,yr^{-1}$ for typical cloud temperatures. The inside-out core collapse model is attractive although clearly has its limitations. For one thing most workers in the field of star formation consider uniform accretion rates unlikely. Nevertheless, the Shu hypothesis seems to be a good starting point.

As the molecular core continues to contract, the density at the center becomes high enough $(10^{-13}\,g\,cm^{-3})$ for it to become optically thick. The temperature at the center then rises for the first time and further contraction is resisted. The first "hydrostatic" core is born.

In the next phase, the temperature and density continues to increase. When temperatures reach around 2,000 K hydrogen molecules are dissociated. Gravitational potential energy is readily absorbed once again and the second collapse phase begins. Temperatures continue to rise and so does the opacity of the gas as hydrogen is ionized. Eventually a second "hydrostatic" core is formed when $\rho \approx 10^{-2}\,g\,cm^{-3}$.

At this stage only a small fraction of the final mass is in the core (at most a few percent) and the main accretion phase begins. The latter lasts for around 10^5 yrs until the star becomes optically visible below the birthline (a concept explained later) on the HR diagram.

Of course the infall models we have been considering up to now are a gross simplification of what we expect to happen in reality. We have ignored the effects of magnetic fields and rotation for example. The net effect in both cases is of course to break the assumed spherical symmetry in favour of axisymmetry. Magnetic fields may help slow down the collapse but ambipolar diffusion ensures that the field diffuses away and gravity wins out.

4 The Main Accretion Phase

Once the hot core has formed, the protostar continues to accrete and builds up most of its mass during a period lasting approximately 10^5 years for a star that eventually ends up with a mass of around 1 $M_\odot$ (see Fig. 4). This is the so-called Class 0 phase; the object is now a true protostar. Class 0 stars are rare (as the phase lasts such a short time) and have a characteristic black-body spectrum [2]. Millimeter waves, emitted by the circumstellar material,

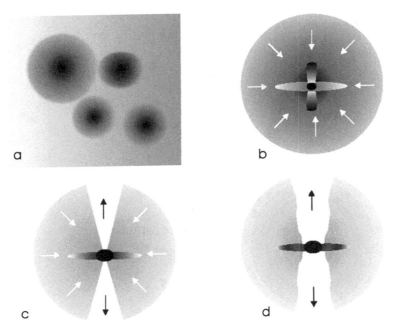

Fig. 4. Diagrammatic illustration of the different phases in the formation of a star. (a) Pre-stellar isothermal molecular cores with density increasing towards the center. (b) Inside-out collapse occurs, an accretion disk develops and an associated outflow. Much of the mass however is still external to the hydrostatic core. This is the Class 0 phase. (c) Most of the mass is now in the center, the disk continues to accrete from the envelope and the star from the disk. The outflow is now well developed and the star has reached the Class I phase. (d) Much of the envelope has now been accreted, or dispersed by the outflow. The star is within a few per cent of its final mass and is optically visible either as a T Tauri star or a Herbig Ae/Be star. This is the Class II phase

can easily penetrate the surrounding disk and envelope, i.e. the circumstellar environment is optically thin in the millimeter band. Thus the miilimeter emission can be used to measure the total mass in the envelope and disk. A Class O protostar is defined as one having more than 50% of its final mass in its disk and envelope. The Class 0 phase is clearly optically invisible. Instead they emit most of their radiation in the far-infrared to millimeter bands. Already, at the Class 0 phase, outflows from the young star have begun (see Sect. 6); outflows that can be seen, for example, through doppler-shifted CO rotational line emission. During this phase the accretion is not expected to be steady (as predicted from simple infall models) but instead to be time varying. How the accretion rate varies with time however is a matter of considerable controversy [20].

The protostar next evolves to the Class I phase; the young star is still surrounded by a disk and envelope [18] although most of the mass is now in the core. The star itself is not directly optically visible although scattered light might be seen in an associated nebula. Instead the peak of its emission is in the

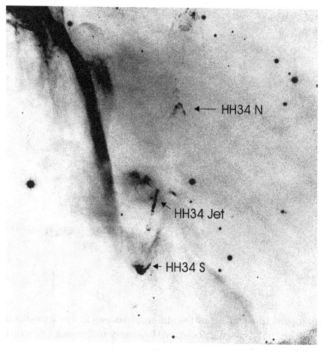

Fig. 5. An emission line image of the HH 34 outflow. HH 34S is the bow-shaped object to the bottom (*south*) with the HH 34 blue-shifted jet pointing towards it. It was thought at one time that HH 34S is the "terminal shock" were the outflow from the young star rams into the surrounding medium. We now know this simple idea is incorrect as the outflow is much bigger than previously thought. A counter bow, HH 34N, is observed to the top (*north*) although no counter-jet is seen. Additional HH emission can be seen further north. Image courtesy of Jochen Eislöffel

near-infrared with a long tail extending to the far-infrared and even millimeter wavelengths (see Fig. 5). Such systems are modelled by a combination of a star (typically with a surface temperature of a few thousand degrees Kelvin) embedded in an optically thick envelope and surrounded by a disk. Much of the accretion is thought to occur onto the disk from the envelope and from there onto the star. As matter grinds its way through the disk towards the star, viscosity ensures that the disk is heated. This heat is radiated away giving rise to the characteristic spectrum observed. Class I protostars are more common than those of Class 0; presumably this reflects the fact that a young star typically spends a longer time in the Class I phase than in Class 0. When modelling Class I protostars, it is normally assumed that the envelope is partially evacuated, above and below the disk, by the outflow. Lines of sight through these bipolar cavities are clearly less extincted and thus a Class I protostar observed at certain angles, may show an enhanced contribution from the star with respect to the envelope and disk.

While the star goes through the Class 0 and I phases; the temperature at the core becomes high enough to ignite deuterium. Deuterium burning is very efficient (and in fact the energy output from deuterium burning is comparable to what the star derives from gravitational contraction). As the protostar is also fully convective, any deuterium that is accreted onto the young star is burnt relatively quickly. Since the process has clear parallels with what happens when a star is on the main sequence, the protostars essentially join an equivalent line in the HR diagram above the main sequence, this is the so-called birthline [49]. Once the main accretion phase is over, and the star becomes optically visible for the first time, deuterium buring ceases and the young star descends from the birthline. The final stages of stellar birth has begun.

5 The Objects of Joy

As mentioned in the Introduction, Joy discovered the T Tauri class of variables in 1945 [21]. At the time, it was not clear to him what they were although their light curves were clearly very irregular with no obvious periodicity. Closer study over the following decades showed us that Ambartsumian's suggestion [1] was correct, i.e. the T Tauri stars were pre-main sequence stars, shining through gravitational contraction as they make their way down the HR diagram. Apart from an irregular light-curve, it was also soon realized that T Tauri stars have a rather unusual spectrum. Unlike typical main sequence stars, $H\alpha$ is strongly seen in emission (occasionally the $H\alpha$ equivalent width is 200–300 Å) and a number of forbidden emission lines such as those due to transitions of OI, SII and NII (e.g. [10]). Generally speaking, these emission lines were found to be blueshifted, a feature that was at first found to be very puzzling. Appenzeller and his colleagues [3] then came up with the novel suggestion that the redshifted component is obscured by a disk. Early researchers in this area had attempted to model the emission lines in terms of a low density stellar wind. The models themselves were not very successful, in particular they produced poor fits to the line profiles (e.g. [19]).

The resolution of this problem came with closer study of the forbidden line emission: it was realised that two or more components are often present (e.g. [31]). The high velocity component normally has a velocity of a few hundred km s^{-1} and is often seen to be spatially extended using long-slit spectroscopy. In contrast the low velocity component, with velocities closer to that of the star, is frequently associated with a denser region and is quite compact. It was suggested [35] that the low velocity component is a disk wind launched at a few AU from the star. The high velocity component, in contrast, they suggested comes from a region much closer to the source and is, in fact a jet at the star. The jet may, or may not, manifest itself on extended scales, in the form of an outflow (see below and the chapter by S. Cabrit).

6 Outflows

The discovery of outflows from young stars can be traced back to the pioneering work of George Herbig [27, 28] and Guillermo Haro [23, 24] who independently discovered the nebulous patches in the sky that bear their names. That these nebulae were linked to the star formation process was immediately obvious; however the precise nature of the link remained a mystery for many years. Richard Schwartz [46] was the first to notice that the emission line spectra of Herbig-Haro (HH) objects resembles those of supernova remnants suggesting that their emission arises from post-shock cooling rather than photo-ionization. Other differences from photo-ionised regions were also evident: for example, HH objects were observed to move at speeds of up to several hundred km s^{-1}, i.e. much faster than the expansion velocities of HII regions. On the other hand the velocities of HH objects were still typically lower than that of the average supernova remnant [16, 30]. Moreover their spectra were dominated by low excitation species (e.g. O, S$^+$) suggesting low velocity shocks.

Several theories were put forward to explain their origin. Richard Schwartz [47], for example, suggested HH objects were due to clumps, from the parent molecular cloud, getting in the way of the wind from a young star. It was however the widespread use of the CCD camera, which allowed much deeper images of the sky than traditional photographic plates, that revealed their true nature. It was found (e.g. [40]) that HH objects were either parts of jets or delineated the region where the jet interacted with its surroundings. The archetypal example, HH 34, is shown in Fig. 5. The first photographic observations of HH 34 [29] showed it to be bow-shaped. Deeper CCD images e.g. [14, 45] however revealed a jet pointing towards HH 34 as well as a counter bow to the north (HH 34N). Close study revealed the jet to consist of a string of HH knots and its source, HH 34-IRS, was only seen in the infrared. The discovery of other jets followed, e.g. HH 111 [25] but the basic structure was usually the same: a string of knots close to the source, i.e. the jet, followed by a more widely spaced set of HH complexes.

As the format of the CCDs became larger and larger so, with time, did the apparent size of HH outflows. Apparent size seemed to be a function of observer patience, allocated observing time and detector field of view! (e.g. [7, 17, 44]). A number of outflows are now known to stretch for at least ten or more parsecs. No doubt, in some cases, they go beyond the parent molecular cloud boundaries. In retrospect this seems reasonable: statistically the outflow phase lasts about 10^6 yrs for low mass stars. If we assume the head of the outflow moves outwards at 50–100 km s^{-1} over this timescale, then clearly it will penetrate many parsecs into its surroundings.

Prior to the discovery of HH jets, it was already evident that YSOs had outflows. Winds from young stars, for example had been discovered many years before [34] but perhaps more pertinently observations of star forming regions using molecular lines (e.g. CO rotational transitions) at mm wavelengths revealed giant redshifted and blueshifted lobes straddling opposite sides of protostars. Some 60 of these "molecular outflows" had been discovered by the mid 80's alone [36]. In all cases they appeared to be poorly collimated with typical velocities of tens of km s^{-1} and sizes between 0.1 – 1 pc (e.g. [6]). This molecular gas does not appear to be ejected directly from the young star but instead is ambient material that is either entrained or pushed by its underlying, much more highly collimated, outflow.

In contrast to Active Galactic Nuclei (AGN) jets, the detailed physics of which is poorly known, the rich emission line spectra of HH jets provide us with a wealth of information on conditions in both the jet propagation and launching zones (see the chapter by Cabrit in this volume for more details of the jet launching zone). For example, we can determine their velocities, from spectroscopic and proper motion studies, to be typically a few hundred km s^{-1} and, as we know their average temperatures are around 10^4K from excitation levels, it follows that jet Mach numbers are 10–30. Another interesting observational finding is that their opening angles, defined as jet width divided by distance from the source, are very small (see chapter by Cabrit). Typical values are a few degrees at most, implying HH jets are highly collimated.

Most of the knots in HH jets have line spectra indicative of shock velocities around 40–80 km s^{-1}. This is considerably lower than actual jet velocities and immediately implies that the knots are not "terminal shocks". Instead it seems much more likely that they are due to variations in the outflow from the source (see, for example, [42]). The basic idea is rather simple: a highly supersonic flow will produce internal shocks (also known as internal "working surfaces") if it undergoes variations in *relative* velocities that exceed the local sound speed. To give a concrete example if the outflow from a source increases in velocity from say 200 km s^{-1} to 230 km s^{-1}, this produces two shocks within the system: the first where the slower jet material ahead of the variation is accelerated and a second where the higher velocity material is decelerated. The material between the two shocks is of course heated and hence under pressure. In the case of a 3-dimensional jet this material is then ejected sideways which, after interacting with its surroundings, gives HH jet knots their characteristic bow-shape.

A major difference between AGN and YSO jets (apart from the obvious ones of size and velocities) is that the former are essentially adiabatic whereas energy losses, due to radiative cooling, are important in the latter. That said, it is still thought that the knots seen in AGN jets have an indentical origin: i.e. they are internal working surfaces. Moreover like AGN jets, YSO jets are seen to curve gently in C- or S-shaped morphologies centred on their source. It seems likely that the causes are again identical: a "side-wind" generated by the motion of the source and precession respectively (e.g. [8]).

The strength of the line emission, in conjunction with abundance estimates and known distances, can be used to gauge the mass loss rates in HH jets. Although there are considerable variations, it is found that these are around 10^{-9} to 10^{-7} $M_\odot$ yr^{-1} for solar mass YSOs [5]. In turn these values can be used, in conjunction with known accretion rates [22] for the same stars, to estimate the outflow efficiency, i.e. the mass loss rate over the accretion rate. Typically this is found to be 1–10%. Such relatively high efficiencies seem to support the idea that jets are launched centrifugally along magnetic field lines anchored to the star's accretion disk (see the chapter by Cabrit for further details). Finally it is worth noting that recent observations, with the long-slit spectrograph on board the HST, have shown evidence that HH jets may be rotating [4, 15, 51]. This is illustrated in Fig. 6. The velocity differences

Jet rotation with 'parallel' slits: DG Tau Jet
Observed Radial Velocity Shift

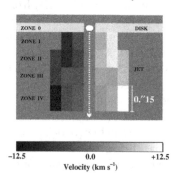

Jet rotation with 'perpendicular' slits: TH 28 jet

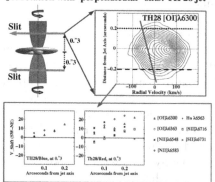

Fig. 6. The Space Telescope Imaging Spectrograph (STIS) was used to measure velocity differences in emission lines *across* jets close to their source (at angular distances of a fraction of an arcsecond). Here the case of DG Tau (*Left*) and Th28 (*Right*) are illustrated. Relative shifts were measured using gaussian fitting and cross-correlation to the line profiles. Measured shifts are around 5–25 km s^{-1}. Such values suggest the jet might be rotating at speeds of 10–20 km s^{-1} at its boundaries. From [43]

observed, on opposite sides of a jet close to its source, are also consistent with models that suggest jets are launched centrifugally from a disk (see also the chapter by Ferreira).

7 Disks Around YSOs

We started this introduction to star formation by mentioning the nebular hypothesis of Kant and Laplace. It is appropriate to end then with the modern day vindication of their ideas on how stars and planets form. As already mentioned there is plenty of indirect evidence for disks around young stars including infrared/mm excesses and the blocking of their redshifted emission lines. Because of their small angular size however direct imaging of disks had to await HST (e.g. [41]) and long baseline millimeter interferometry (e.g. [33]). Such studies show that the disks surrounding T Tauri stars and other low mass stars in irradiated regions such as Orion, have diameters around 100 AU.

Perhaps the most dramatic form in which we see disks is the so-called proplyds. We have already mentioned the cluster at the centre of the Orion

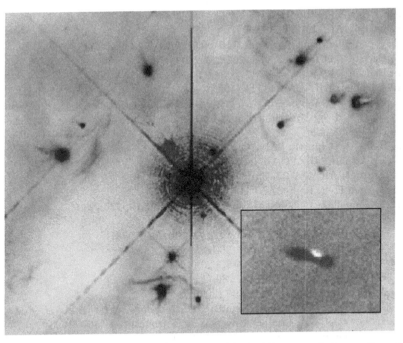

Fig. 7. HST images of proplyds in the Orion Nebula. These proplyds are all centred on θ^1C Orionis and the image is courtesy of John Bally, David Devine and Ralph Sutherland. The bow shocks are regions where photo-evaporated gas from the disks interacting with the wind from this O-type star. Inset a disk is seen in silhouette against the emission from the Orion Nebula. This YSO is outside the HII region. Note the scattered light from the central YSO

nebula, the Orion Cluster (see Fig. 1). It consists of approximately 700 stars, many of which are contained within the HII region ionized, primarily, by the most powerful of the Trapezium stars, θ^1C Orionis. HST showed that many of these low mass stars are surrounded by extended structures (see Fig. 7) with typical angular dimensions of about 1″. Some have cometary-like tails pointing away from θ^1C Orionis. It was also shown that protostars are detected in the near-infrared at the center of almost all of these objects [38]. We had finally seeing Kant and Laplace's disks.

Acknowledgements

TPR wishes to thank the organisers of the Villiard de Lans JETSET School for all their hard work and the great hospitality he was shown while there.

References

1. V. A. Ambartsumian, In *Stellar Evolution and Astrophysics*, (Erevan: Acad. Sci. Armenian, SSR) (1947)
2. P. André: IAUS **182**, 483 (1997)
3. I. Appenzeller, I. Jankovics, R. Östreicher: A&A **141**, 108 (1984)
4. F. Bacciotti, et al.: ApJ **576**, 222 (2002)
5. F. Bacciotti, et al.: ApSS **287**, 3 (2003)
6. R. Bachiller: ARA&A **34**, 111 (1996)
7. J. Bally, J., D. Devine: In *IAU Symp. 182: Herbig-Haro Flows and the Birth of Stars*, (ed. by B. Reipurth and C. Bertout) Kluwer Academic Publishers, p. 29 (1997)
8. J. Bally, B. Reipurth: RMxAC **13**, 1 (2002)
9. E. E. Barnard: JRASC **10**, 241 (1916)
10. C. Bertout: ARA&A **27**, 351 (1989)
11. B. J. Bok, E. F. Reilly: ApJ **105**, 255 (1947)
12. I. A. Bonnell, S. G. Vine, M. R. Bate: MNRAS **349**, 735 (2004)
13. T. L. Bourke, et al.: ApJ **554**, 916 (2001)
14. T. Bührke, R. Mundt, T. P. Ray: A&A, **200**, 99 (1988)
15. D. Coffey, et al.: ApJ **604**, 758 (2004)
16. M. Cohen, G. A. Fuller: ApJ **296**, 620 (1985)
17. J. Eislöffel, R. Mundt: AJ **114**, 280 (1997)
18. J. A. Eisner, et al.: ApJ **635**, 396 (2005)
19. S. Edwards, T. P. Ray, R. Mundt: In: *Protostars and Planets III*, (ed. by E. H. Levy and J. I. Lunine) Unversity of Arizona Press, 567 (1993)
20. D. Froebrich, et al.: MNRAS **368**, 435 (2006)
21. A. H. Joy: ApJ **102**, 168 (1945)
22. P. Hartigan, S. Edwards, L. Ghandour: ApJ. **452**, 736 (1995)
23. G. Haro: ApJ **115**, 572 (1952)
24. G. Haro: ApJ **117**, 73 (1953)
25. P. Hartigan, et al.: ApJ **559**, L157 (2001)

26. C. Hayashi: ARA&A **4**, 171 (1966)
27. G. H. Herbig: ApJ **111**, 11 (1950)
28. G. H. Herbig: ApJ **113**, 697 (1951)
29. G. H. Herbig: Lick Observatory Bulletin **658**, 1 (1974)
30. G. H. Herbig, B. F. Jones: AJ **86**, 1232 (1981)
31. G. A. Hirth, R. Mundt, J. Solf: A&AS **126**, 437 (1997)
32. M. Houde et al.: ApJ **604**, 717 (2004)
33. Y. Kitamura, R. Kawabe, M. Saito: ApJ **465**, L137 (1996)
34. L. V. Kuhi: ApJ **140**, 1409 (1964)
35. J. Kwan, E. Tademaru: ApJ **454**, 382 (1995)
36. C. J. Lada: ARA&A **23**, 267 (1985)
37. R. B. Larson: MNRAS **194**, 809 (1981)
38. M. J., McCaughrean, J. R. Stauffer: AJ **108**, 1382 (1994)
39. L. Mestel, L. Spitzer: MNRAS **116**, 503 (1956)
40. R. Mundt, E. W. Brugel, T. Bührke: ApJ **319**, 275 (1987)
41. C. R. O'Dell: AJ **115**, 263 (1998)
42. A. C. Raga, T. Beck, A. Riera: ApSS **293**, 27 (2004)
43. T. P. Ray, C. Dougados, F. Bacciotti, J. Eislöffel, A. Chrysostomou: Toward resolving the outflow engine: An observational perspective. In: *Protostars and Planets V*, (ed. by B. Reipurth, D. Jewitt, and K. Keil) University of Arizona Press, Tucson, pp. 231–244 (2007)
44. T. P. Ray: A&A **171**, 145 (1987)
45. B. Reipurth, et al.: A&A **164**, 51 (1986)
46. R. D. Schwartz: ApJ **195**, 631 (1975)
47. R. D. Schwartz: ApJ **223**, 884 (1978)
48. F. H. Shu: ApJ **214**, 488 (1977)
49. S. W. Stahler: ApJ **332**, 804 (1988)
50. L. Testi, A. I. Sargent: ApJ **508**, L91 (1998)
51. J. Woitas, et al.: A&A **432**, 149 (2005)
52. M. Wolf: AN **219**, 109 (1923)

Jets from Young Stars: The Need for MHD Collimation and Acceleration Processes

Sylvie Cabrit

LERMA, Observatoire de Paris, CNRS, 61 Av. de l'Observatoire, 75014 Paris, France
sylvie.cabrit@obspm.fr

Abstract. This lecture revisits in the light of recent data the main lines of evidence indicating that MHD processes play a crucial role in jets from young stars. Measurements of jet collimation and jet ejection-accretion efficiencies are reviewed and compared at various evolutionary stages (from protostars to optically revealed objects). It is then shown that they cannot satisfactorily be accounted for by purely hydrodynamical processes. MHD magneto-centrifugal ejection (combining magnetic self-collimation and magnetic acceleration) appears as the most effective mechanism able to reproduce the observed jet properties at all evolutionary phases. The jets would then be intimately linked to angular momentum extraction from the accreting disk and/or star.

S. Cabrit: *Jets from Young Stars*, Lect. Notes Phys. **723**, 21–53 (2007)
DOI 10.1007/978-3-540-68035-2_2

1 Measurements of Jet Collimation at Various Stages

Highly collimated, energetic jets are observed at all evolutionary stages of low-mass young stellar objects (hereafter YSOs; see Ray, this volume). Their morphological and kinematic properties remain surprisingly similar despite different tracers and circumstellar environments (see e.g. [14] for a comparative review). Scaled-up versions are also encountered in luminous YSOs of several 10^4 $L_\odot$ (e.g. [23, 33, 75]), but their greater distance from us hampers detailed studies on 100 AU scales. In the following, I will thus focus on the collimation properties of nearby jets from low-luminosity sources, which provide the tightest constraints for collimation models.

1.1 Jet Widths and Opening Angles in Class II Sources

In optically visible pre-main sequence stars (Class II sources), small "microjets" are traced out to ≤ 1000 AU of the source in low-excitation forbidden lines in the optical (e.g. [O I]$\lambda6300$, [N II]$\lambda6583$, [S II]$\lambda6731$) and near-IR ([Fe II]$\lambda1.64\mu$m). At this late stage (a few Myr), circumstellar material has settled into a thin disk, allowing a clear line of sight to the innermost jet regions where initial collimation is taking place.

Measurements of jet widths within 800 AU of the star, obtained with HST or with adaptive-optics from the ground, are presented in Fig. 1 (see also [89] and [106]). The jets are well resolved transversally at 30–50 AU from the central star, with a characteristic width of 20–40 AU. Beyond 50 AU, the jet width grows slowly with distance with a full opening angle of a few degrees. This is compatible with the "Mach angle" expected for free lateral expansion of a supersonic jet:

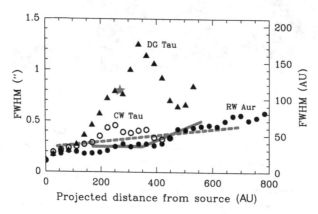

Fig. 1. Optical jet widths within 800 AU for Class II sources. Symbols show adaptive optics observations with CFHT (deconvolved). Solid/dashed curves show HST results for HH 30 and HL Tau from [89]. Measurements for DG Tau are affected by strong bowshocks and are an upper limit to the true jet width. From [38]

$$\alpha = 2 \times \arctan(C_s/V_j) \simeq 3.8^o \qquad (1)$$

for a typical jet temperature of 10^4 K and a jet speed of 300 km s^{-1}.

Within 50 AU (0.3" at the distance of Taurus), accurate correction for the instrumental PSF (0.1" with HST/STIS) becomes critical to retrieve intrinsic jet widths. Remarkable results have been recently obtained by Hartigan et al. (2004) on two Class II jets using STIS/HST in "slitless" imaging mode. Results are shown in Fig. 2. The jet width is again 25–30 AU at 50 AU, but the jet opening angle inside this region is much wider (20°–30°) than on larger scales.

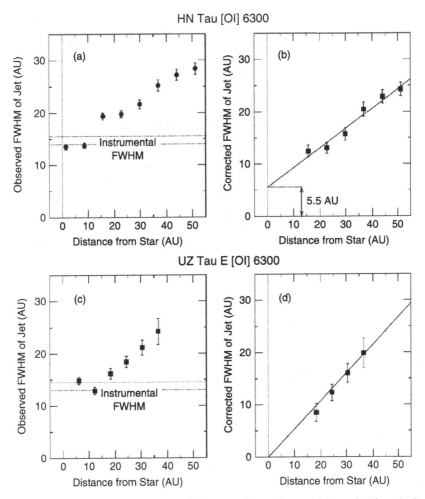

Fig. 2. Optical jet widths within 50 AU in two Class II stars, before (*left*) and after (*right*) correction for the PSF. The full opening angle is 20° in HN Tau (*top*) and 30° in UZ Tau E (*bottom*). From Hartigan et al. (2004). Similar widths are found for bipolar water maser systems in Class 0 sources (see text)

Extrapolation back to the stellar position indicates that the optical jets originate from a region less than 5.5 AU in diameter, and possibly much smaller.

1.2 Jet Widths and Opening Angles in Class I Sources

Atomic jets from 10^5yr old infrared protostars (Class I objects) are brighter than in the later Class II phase, with a continuous beam that may be traced in the optical out to 0.05 pc from the source (see e.g. [90] for a review). Due to high extinction and scattered light nebulosity, the innermost jet regions are only traced in the near-IR, using H_2 and [Fe II] lines [30, 31, 32].

The opening angles and widths of Class I jets agree with an extrapolation of the Class II jet properties in Fig. 1. For example, the HH 1 jet has a width of 100 AU at 1000 AU from the source, with a full opening angle of 2.6° thereafter [92]. The jet widths appear similar in H_2 2.12μm and in [Fe II], at the resolution of HST [92]. No width measurements have yet been reported for inner jet regions within 1000 AU.

1.3 Jet Widths and Opening Angles in Class 0 Sources

Energetic collimated jets are already present at the earliest phase of protostellar evolution, in the so-called "Class 0" phase where the infalling envelope mass well exceeds the mass of the accreting star (see e.g. the review by [1]). The jet base is heavily extincted, but near-infrared H_2 and [Fe II]$\lambda1.64\mu$m emission knots are seen beyond a few hundreds of AU from the source (cf. [20] and references therein). Spectacular CO or SiO jets have also been recently discovered in a few Class 0 sources with mm interferometers. L 1488 and HH 211 are good examples [24, 53, 54]. On resolved scales (beyond 200 AU) the jet width, opening angle, and even knot spacing, are strikingly similar to those of atomic Class I jets such as HH 34 [14].

The innermost jet regions in Class 0 objects have first been imaged in the thermal free-free continuum at centimeter wavelengths with the VLA (see e.g. [2] and [94] for reviews). They are unresolved transversally with a maximum width < 50 AU on 50 AU scale.

New clues to the small scale collimation of Class 0 jets have been recently provided by VLBI-cm studies of H_2O masers. Forty percent of Class 0 objects excite water masers tracing very dense shocks within 100 AU of the source ([50] and references therein). When mapped, the maser spots often display a bipolar structure with proper motions indicative of a jet-like flow, with a diameter of only 8–20 AU at a distance of 20–40 AU [22, 25, 49]. This is strikingly close to optical jet widths measured on the same scale in Class II sources (cf. Fig. 2).

1.4 Jet-disk Alignment and Jet Precession

In Class II objects where the disk has been imaged in scattered light or CO lines, one may compare the direction of the optical jet with the symmetry axis of the disk. In all cases studied so far, the two axes coincide precisely, even when the jet curves away from its initial direction further out. A fine example is the HH 30 system, whose nearly edge-on disk and associated jet were discussed by Burrows et al. (1996). Other examples are HV Tau/C [102] and RW Aur [19].

Observations further show that the jet axis varies by at most a few degrees within 500 AU of the source, corresponding to timescales ≤ 10 yrs. For example, Raga et al. (2001) modelled the small wiggles in the DG Tau microjet with a precession angle of about 5° over an 8 yr period. More pronounced jet wiggles and jet curving are observed on larger scale in Class I jets (cf. [91]) and in parsec-scale chains of emission knots (e.g. [36]). But the typical timescales are $10^2 - 10^4$ yrs, much longer than dynamical timescales in the jet ejection zone.

1.5 Summary

The measurements described above yield several important implications and constraints for the collimation process of jets from young stars.

- Studies of Class II stars show that the optical jet originates from a region less than 3 AU in radius and expands with a full opening angle of about 20°–30° between 10 and 50 AU. Beyond 50 AU, the jet opening angle is much smaller (a few degrees) and compatible with free supersonic expansion. Hence, the collimation process must enforce an essentially unidirectional flow at about 50 AU of the source, at least for the *brightest part* of the jet.
- Quite strikingly, jet widths are not found to differ between objects of varied evolutionary stages, from Class 0 protostars with a dense infalling envelope, through Class I infrared sources with residual infall, to Class II optically visible stars with no envelope and a thin disk. Thus, jet collimation cannot rely on an infalling envelope.
- The collimation process must produce a jet beam that is closely aligned with the disk axis at 30–50 AU from the source. Jet precession does not exceed a few degrees on timescales ≤ 10 yrs.

2 Comparison with HD and MHD Jet Collimation Models

In this section I will revisit (in chronological order) the various jet collimation models that have been proposed in the literature, and examine whether they can reproduce recent data on jet widths and collimation scales. I will focus on the case of low-luminosity Class II sources (classical T Tauri stars), which provide the most severe constraints. For quantitative estimates, a fiducial mass-loss rate of $10^{-8} M_\odot \mathrm{yr}^{-1}$ will be adopted, as this is typical of Class II jets where collimation measurements have been made.

2.1 Collimation by External Thermal Pressure

The first scenario proposed to explain the collimation of YSO outflows involved the external collimation of an initially isotropic wind by a flattened circumstellar structure, e.g. a disk. The anisotropic thermal pressure gradient limits the wind expansion in the disk plane, forcing it to collimate towards the disk poles.

When the wind ram-pressure is exactly balanced by the ambient thermal pressure, the cavity reaches a steady configuration and does not expand any more. Semi-analytical calculations of the steady cavity shape were performed in two extreme assumptions concerning the wind shock: adiabatic [67] and highly radiative [8, 101]. The latter hypothesis is probably more correct for the high densities and moderate speeds of YSO winds [41].

Figure 3 represents the steady cavity shape found by Barral and Canto (1981) for a disk immersed in a constant pressure cloud. As the wind shock is assumed highly radiative, the shocked wind material is compressed in a thin layer and slides against the cavity wall (the momentum parallel to the shock is conserved). Below we discuss in turn three characteristics of the shocked wind cavity: (1) its "waist" radius in the equatorial disk plane, (2) its asymptotic opening angle above the disk plane, and (3) the distance Z_{max} where the wind is refocussed on-axis under the effect of a finite ambient pressure.

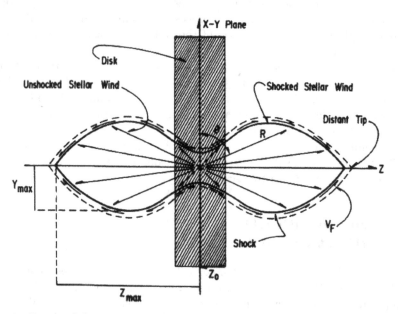

Fig. 3. Sketch of the geometry of the shocked wind cavity for by a uniform self-gravitating disk (*shaded area*), embedded in a finite pressure medium. Ambient pressure forces the shocked wind to refocus on axis at the "tips" of the cavity. From [8]

1- Equatorial Confinement ("Waist" Radius)

In the simple case of an initially isotropic wind, the equatorial radius R_0 of the "waist" of the cavity will be determined by balance between the wind ram pressure and the thermal pressure in the disk plane,

$$\rho_w(R_0)V_w^2 = \dot{M}_w V_w/4\pi R_0^2 = P_c(R_0) , \qquad (2)$$

where $\dot{M}_w$ is the total wind mass-loss rate. Denoting as T_c and n_c the temperature and number density in the disk plane, one finds:

$$\left(\frac{\dot{M}_w}{10^{-8}M_\odot \mathrm{yr}^{-1}}\right)\left(\frac{V_w}{300\mathrm{km\ s}^{-1}}\right)\left(\frac{R_0}{100\mathrm{AU}}\right)^{-2} = \left(\frac{n_c(R_0)T_c(R_0)}{6\times 10^9\mathrm{K\ cm}^{-3}}\right) . \qquad (3)$$

Midplane densities and temperatures in Class II sources have been recently estimated at radii of 50–400 AU from resolved interferometric disk maps in CO lines and dust thermal continuum. Detailed analyses exist for a few well-studied disks, e.g. GM Aur and DM Tau [29, 40], and they suggest $n_c(r)T_c(r) \simeq 1 - 2\times 10^{10}(r/100\ \mathrm{AU})^{-3.4}$ K cm^{-3}, for an accretion rate $\dot{M}_{acc} \simeq 10^{-8}$ $M_\odot \mathrm{yr}^{-1}$ [55].

We may also calculate theoretically $n_c(r)T_c(r)$ for a "standard" steady viscous accretion disk model where viscosity scales as $\nu = \alpha C_s h$ (with h the local disk scale height) and α is assumed constant throughout the disk. In a thin and vertically isothermal disk dominated by gravity of the central star, $h/r = C_s/V_K(r)$ and $\rho_c = \Sigma/(\sqrt{2\pi}h)$ (gaussian profile) where Σ is the surface density. The steady inward accretion speed is $u_R \simeq \nu/r$ [72]. Therefore,

$$\rho_c C_s^2 = \frac{\dot{M}_{acc}C_s^2}{2\pi r h u_R\sqrt{2\pi}} = \frac{1}{\alpha(h/r)}\left(\frac{\dot{M}_{acc}V_K}{(2\pi)^{3/2}r^2}\right) . \qquad (4)$$

With $h/r \simeq 0.1$ and $\alpha \simeq 10^{-2}$, as suggested by disk images and disk lifetimes in T Tauri stars (e.g. [13, 60]), we obtain:

$$n_c(r)T_c(r) \simeq 5\times 10^{10}\left(\frac{r}{100\ \mathrm{AU}}\right)^{-5/2}\left(\frac{\dot{M}_{acc}}{10^{-8}M_\odot \mathrm{yr}^{-1}}\right)^{1/2}\left(\frac{M_\star}{M_\odot}\right)^{1/4}\mathrm{K\ cm}^{-3} .$$
$$(5)$$

This is similar to observationally determined values at 100 AU, although the index of the radial power law is flatter (-2.5 instead of -3.4).

Comparing with (3), we see that the midplane pressure of an α-disk will exceed the wind ram pressure already at disk radii of 100 AU, especially since sources with mass-loss rate $\dot{M}_w = 10^{-8}$ $M_\odot \mathrm{yr}^{-1}$ have on average 10 times higher mass-accretion rate (cf. Sect. 3.1). As $n_c T_c$ increases inward more steeply than r^{-2}, equatorial wind expansion will in fact be stopped at the inner edge of the disk, so that $R_0 \leq 0.1$ AU, consistent with the constraint $R_0 \leq 3$ AU obtained by Hartigan et al. (see Fig. 2).

Thus, Class II viscous accretion disks provide very efficient *equatorial* wind confinement, even for wind mass-loss rates comparable to the disk accretion rate. However, confinement in the equatorial plane is not sufficient to produce a jet. The opening angle of the shocked wind cavity should also be small enough. We now examine this issue.

2- Wind Opening Angle

Simple geometrical arguments show that the asymptotic opening angle of the shocked wind cavity (above the disk plane) will depend on the *ratio* of the disk pressure scale-height h to the waist radius R_0, i.e. on the aspect angle h/R_0: A larger h/R_0 means that a wind streamline flowing at a given angle to the midplane will encounter a higher ambient pressure, hence experience a more efficient confinement, yielding a narrower cavity.

The asymptotic opening angle of the cavity has been calculated in the particular case of a uniform, isothermal, self-gravitating disk ($h = \mathrm{cst}$) by Barral and Cantó (1981). As illustrated in Fig. 4, the cavity angle *from the disk midplane*, θ^*, depends very strongly on $\lambda = h/R_0$. Cylindrical collimation ($\theta^* = 90°$) is achieved for $\lambda = 0.8$.

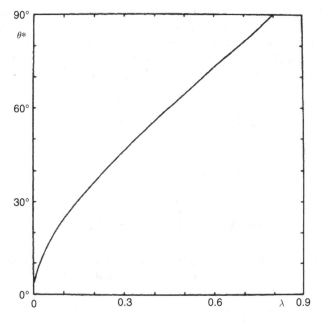

Fig. 4. Asymptotic angle made by the wind cavity with *the disk midplane*, θ^*, as a function of the disk aspect ratio at the waist radius, $\lambda = h/R_0$, for a self-gravitating isothermal disk. Cylindrical collimation ($\theta^* = 90°$) is achieved for $h/R_0 = 0.8$. From [8]

While such large values of h/R_0 could be expected in self-gravitating disks around young objects (see (13) of [8]), disks around Class II sources are dominated by the stellar gravity and have typically $h/r = C_s/V_K \leq 0.1$ at all radii. From Fig. 4, this would suggest an asymptotic angle $\theta^* \simeq 25°$ from the equator, ie a full cavity opening angle of 130°. Note that this estimate is too optimistic, as Fig. 4 was derived for self-gravitating disks, where pressure decreases vertically as $\exp(-2z/h)$, whereas disks dominated by gravity from the central star have a steeper decrease, $\propto \exp(-(z/h)^2/2)$. Thus, disk thermal pressure alone will not produce a well-focussed jet beam in Class II sources.

3 - Reconfinement by Ambient Thermal Pressure

Barral and Cantó (1981) showed that strong reconfinement of the shocked wind can be obtained if the disk is immersed in a medium of non-zero pressure. Since the wind pressure drops as R^{-2}, the (constant) ambient pressure eventually dominates and forces the shocked wind to refocus towards the axis, forming an acute tip at a distance Z_{max} from the star, as illustrated in Fig. 3. A narrow, supersonic jet parallel to the disk axis then emerges from this tip [105].

Unfortunately, the observational constraint that $Z_{max} \leq 50$AU turns out to require excessively high ambient densities: Careful examination of Fig. 5 of Barral and Cantó (1981) shows that, for all values of Z_0/h (where Z_0 is the altitude at which pressure becomes constant), Z_{max} obeys

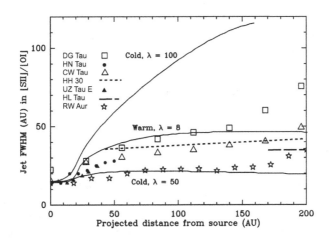

Fig. 5. Comparison of observed optical jet widths in Class II sources (*symbol meaning indicated in the upper left*) with predicted widths for self-similar MHD disk-wind solutions launched from 0.07–1 AU (*solid curves*). Model predictions are obtained from synthetic emission maps in the same emission lines and at the same resolution (15 AU) as the observations. From [89]

$$Z_{max}^2 \simeq R_0^2 (P_c/P_0) = \dot{M}_w V_w/(4\pi P_0) . \qquad (6)$$

This says, not too surprisingly, that the ambient pressure P_0 must be on the order of the wind ram pressure at Z_{max} for reconfinement to occur. Hence, using (3) above, we see that reconfinement at distances ≤ 50 AU of the central star would require $(n_{coll} \times T) \geq 2.4 \times 10^{10}$ K cm^{-3} for a wind mass-loss rate of 10^{-8} $M_\odot$yr^{-1}.

Assuming the collimating material is at typical cloud temperatures $\simeq 10$ K, it would produce a large visual extinction towards the central source $A_V \geq n_{coll} Z_{max}/(10^{21} \text{cm}^{-2}) \geq 1800(10$ K$/T)$ mag. However, T Tauri stars suffer typically less than 3 magnitudes of extinction (see e.g. [58]), hence the density of cold remnant gas fails by at least a factor 600 to collimate the jets. Note that wind refocussing could be achieved in dense infalling envelopes around younger protostellar sources, as nicely demonstrated in the numerical simulations of Delamarter, Frank, and Hartmann (2000). However, we have seen that jets in Class II objects are as well collimated as those in Class I/0 sources (see previous section), so cold infalling envelopes cannot be the main agent for their collimation.

Returning to the above expression for A_V, we see that external collimation of Class II jets with $A_V < 3$mag would require a *warm* circumstellar medium with $T \geq 6000$ K. The only obvious means to fill a 100 AU region with warm gas around low-mass Class II sources would be through photoevaporation of the UV-irradiated disk surface. However, for typical T Tauri parameters, the predicted density of the photoevaporated disk wind at a distance of 50 AU is only $\simeq 400$cm^{-3} [48]. Even if the flow keeps its initial temperature of 10^4 K despite the expansion, its thermal pressure will be a factor 6000 too small for jet refocussing.

We conclude that external thermal pressure can be safely ruled out as the generic jet collimation process in YSOs.

2.2 Collimation by External Magnetic Pressure

Since interstellar clouds are magnetized, and jet sources tend to have their disk axes aligned with the field [79], we will investigate now whether jet focussing could be provided by ambient magnetic fields. When the circumstellar B-field has a preferred direction, magnetic tension counteracts wind expansion perpendicular to the field lines, leading to a wind bubble elongated along $\boldsymbol{B}$. This process was first investigated in the case of an adiabatic wind shock, analytically by Koenigl (1982) and numerically by Stone and Norman (1992). It was reconsidered in the radiative case, specifically in the context of the collimation of stellar winds from T Tauri stars, by Kwan & Tademaru (1988).

Under radiative shock conditions, an order of magnitude estimate for the collimating field required to focus the wind into a jet at distance Z_{max} may be derived in the same way as in the preceding section, simply replacing ambient thermal pressure by magnetic pressure:

$$\dot{M}_w V_w / (4\pi Z_{max}^2) \simeq B_{coll}^2 / 8\pi . \qquad (7)$$

This is only approximate, as expansion of the wind will compress the field and thus modify slightly the equilibrium condition. Other considerations give a similar expression but with the jet radius $r_j \simeq 15$ AU in place of Z_{max} [64], which would be even more constraining. Quantitatively, we obtain

$$B_{coll} \simeq 10 \left(\frac{\dot{M}_w}{10^{-8} M_\odot yr^{-1}} \right)^{1/2} \left(\frac{V_w}{300 km\ s^{-1}} \right)^{1/2} \left(\frac{Z_{max}}{50 AU} \right)^{-1} mG . \qquad (8)$$

This value of B is much higher than the 10–100 μG measured in dense prestellar cores of densities $n_H \simeq 10^4 cm^{-3}$ [79]. In principle, the large-scale field will be amplified by compression during the core contraction phase [9]. However, the poloidal flux required for collimation appears quite large. For jet collimation to be effective, the confining poloidal field must be anchored over a disk region of radius $r_D \gg r_j \simeq 15$ AU [66]. With $r_D \simeq 100$ AU, the required magnetic flux $(\Phi_B)_{coll} = \pi r_D^2 B_{coll}$ is then

$$(\Phi_B)_{coll} = 8 \times 10^{28} \left(\frac{\dot{M}_w}{10^{-8} M_\odot yr^{-1}} \right)^{1/2} \left(\frac{V_w}{300 km\ s^{-1}} \right)^{1/2} \left(\frac{r_D^2 Z_{max}^{-1}}{200 AU} \right) G\ cm^2 .$$
$$(9)$$

This may be compared with the total magnetic flux initially threading the core just before gravitational collapse. For collapse to occur, the ratio of mass to flux must exceed a critical value $(M/\Phi_B)_{crit} = 0.13/\sqrt{G}$ [80]. Hence, for a typical T Tauri star, the initial flux is at most:

$$(\Phi_B)_{init} = 4 \times 10^{30} \left(\frac{M_{init}}{1 M_\odot} \right) \frac{(M/\Phi_B)_{crit}}{(M/\Phi_B)_{init}} G\ cm^{-2} . \qquad (10)$$

We thus find that any primordial poloidal field collimating the jet should have trapped within 100 AU at least 2% of the poloidal flux initially present in the core (up to 20% if the initial mass/flux ratio is 10 times the critical value). This is quite high, given that some of the flux will presumably end up in the star itself, and that field diffusion should be quite effective at the low ionization levels and high densities in disks. Regeneration of poloidal flux by a global disk dynamo also appears unlikely on scales $r_D \geq 100$ AU.

   Another possibility avoiding the flux problem would be to refocus the wind not by an organized poloidal field, but by a "turbulent" magnetic pressure with $< B > \simeq 10$ mG. Indeed, we have seen previously that confinement in the disk midplane is sufficient to initiate preferential wind expansion towards the poles (though with a very wide opening angle). Isotropic turbulent magnetic pressure could then play the role of the constant ambient thermal pressure in the Barral & Cantó models, simply forcing back the shocked wind into a

narrow jet at Z_{max}, aligned with the disk axis. Since thermal pressure around
T Tauri stars fails at least by a factor 600 to refocus the wind, the magnetic
energy density would have to be well above equipartition (plasma $\beta < 1/600$).
It is unclear how such a magnetically dominated turbulent region may be
created and maintained over 100 AU scales.

2.3 MHD Self-collimation

The third collimation mechanism proposed for YSO jets is the process of *MHD
self-collimation*, well covered in several lectures in this volume (Tsinganos,
Sauty, Ferreira, Lery). Very schematically, the basic elements include an MHD
wind launched along an organized magnetic field anchored in a rotating object
(star, disk, infalling envelope), with a non-zero current flowing across the
magnetic surfaces. Beyond the Alfvén point, the inertia of the wind "drags"
behind and winds-up the field lines, creating a strong toroidal B-component.
The $J \times B$ Lorentz force then exerts a confining force towards the axis that
recollimates the magnetic/flow surfaces.

In the case of MHD winds from accretion disks, the optically visible jet
beam may trace only the densest axial regions in a wider flow, with outer
streamlines collimating on much larger scales (hence the term "optical il-
lusion" introduced by Shu et al. 1995; see also [16]). Detailed calculations
of emission maps *and* convolution by the beam are then necessary to compare with observations. Furthermore, the collimation depends strongly on
the mass-loading of the streamlines [44, 87]. As shown in Fig. 5, the observed jet widths, collimation scales, and opening angles, can be very well
reproduced by self-similar disk wind models with reasonable parameters (see
[51, 85, 89]).

Launching steady MHD disk winds from keplerian disks requires poloidal
fields close to equipartition in the disk plane (see [45]). Thanks to the strong
wind magnetic torque, the accretion speed is much faster ($u_R \simeq C_s$) than in
an α-disk, so the density is correspondingly lower. For a disk accretion rate of
10 times our fiducial wind mass-loss rate of $10^{-8} M_\odot \mathrm{yr}^{-1}$, and a self-similar
model, one finds (cf. 16 in Ferreira's lecture, this volume):

$$B_{equ}(R) \simeq 200 \left(\frac{R}{1 \, \mathrm{AU}} \right)^{-5/4} \left(\frac{\dot{M}_{acc}}{10^{-7} M_\odot \mathrm{yr}^{-1}} \right)^{1/2} \left(\frac{M_\star}{M_\odot} \right)^{1/4} \mathrm{mG} \ . \quad (11)$$

The poloidal magnetic flux enclosed within 1 AU is then (Φ_B) $\simeq 10^{26}$ G cm^2.
Self-collimation of MHD winds from the stellar surface would require an even
smaller magnetic flux (e.g. [95]).

Hence, MHD self-collimation of jets is much more "economical" than external magnetic collimation by a large scale poloidal field, in the sense that
the required *poloidal* flux is at least 2 orders of magnitude smaller. The difference arises because self-collimation is achieved by the toroidal field component
created by the wind itself, not by the poloidal component carrying the flux.

The stability of this self-collimation mechanism is an important issue, as strong toroidal fields might be expected to generate destructive non-axisymmetric "kink-modes". Numerical simulations in 3D suggest that self-collimated MHD jets could actually maintain stability beyond the Alfvén surface through self-regulating processes of internal flux redistribution (see [87] for a recent review).

2.4 Summary

- Jet collimation by external thermal pressure gradients is ruled out: disks are too thin to produce a well-collimated jet beam, and dense confining envelopes would produce excessive extinction towards Class II jet sources.
- Jet collimation by external magnetic pressure gradients would require a magnetically dominated region ($\beta < 1/600$) with a poloidal or turbulent field strength $\simeq 10$ mG over at least 100 AU. It is unclear how such a region may be formed and maintained.
- Self-collimation of MHD winds launched from the disk or star produces collimated jets with only a tiny fraction of the magnetic flux required for external magnetic collimation. Furthermore, detailed predictions for self-similar MHD disk winds readily reproduce the jet widths, collimation scales, and opening angles in Class II jets for realistic parameters. Hence this process appears currently as the most promising.

3 Measurements of Jet Energetics: Accretion-ejection Correlations

The second line of evidence for the role of MHD processes in YSO jets derives from measurements of the jet *energetics* (mass, momentum, and kinetic energy flux), compared to the available power in the driving source.

Depending on the evolutionary stage of the source (Class 0, I, or II), various techniques may be used to estimate the jet energetics. Below I will recall the main results, namely that jet energetics at all stages appear tightly correlated with accretion signatures. This strongly argues that the jets are ultimately powered by the release of gravitational energy in the accretion flow feeding the young star. The *ejection efficiencies*, i.e. the ratios of jet mass, momentum, and kinetic energy flux to the accretion rate (or accretion luminosity) will also be reviewed. They will later be compared with predictions of jet acceleration mechanisms.

In the following, I will use the subscript j to refer to one-sided quantities estimated in *one* jet lobe, and the subscript w (as "wind") to refer to quantities integrated over all solid angles of the outflow.

3.1 Ejection to Accretion Ratio in Atomic Jets (Class II/I)

Class II sources are an ideal laboratory to study ejection-accretion correlations for two reasons: (i) the stellar photosphere is visible, so the excess emission produced by disk accretion can be extracted and modelled, giving a measure of the accretion rate $\dot{M}_{\rm acc}$ and of the accretion luminosity $L_{\rm acc} = GM_\star \dot{M}_{\rm acc}/R_\star$. (ii) the jet is observable in optically thin (forbidden) atomic lines, allowing relatively direct estimates of the jet density, radius, speed, and mass-flux rate.

First indications of an ejection-accretion connexion in Class II sources were presented by Cohen et al. (1989), Cabrit et al. (1990), and Corcoran and Ray (1998) who reported a correlation between the [O I]λ6300 luminosity (from the jet) and the infrared excess luminosity (from the disk) in T Tauri stars and their higher mass analogs, the Herbig Ae/Be stars. No such correlation was found with the stellar photospheric luminosity, implying that the ejection process should be ultimately powered by disk accretion. This is confirmed by the absence of any detectable jet in stars without accretion disks ("weak-lined" T Tauri stars).

The ejection-accretion correlation in T Tauri stars was confirmed and quantified by Hartigan et al. (1995; hereafter HEG95), who developed methods to estimate $\dot{M}_{\rm acc}$ and $\dot{M}_j$ from optical spectra. Their results are presented in the left panel of Fig. 6. A correlation is apparent over 3 orders of magnitude in $\dot{M}_{\rm acc}$, with a mean (one-sided) ratio $\dot{M}_j/\dot{M}_{\rm acc} \simeq 0.01$. This is probably a lower limit: In a detailed re-examination of accretion rate derivations, Gullbring et al. (1998, hereafter G98) concluded that $\dot{M}_{\rm acc}$ values from HEG95 were systematically too large by about a factor 10. Stars with revised $\dot{M}_{\rm acc}$

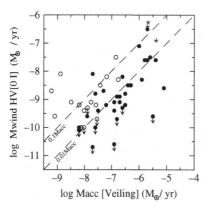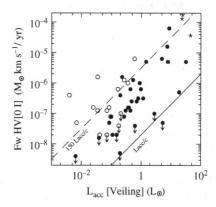

Fig. 6. Left: Mass-loss rate in the blueshifted jet versus disk accretion rate in T Tauri stars. Solid circles = data from [58]. Open circles = revised accretion rates from [55]. Starred symbols = Class I optical jets [57]. Dashed lines plot a (one-sided) ratio $\dot{M}_j/\dot{M}_{\rm acc}$ of 0.01 and 0.1. **Right**: same as left panel, for the momentum flux in the blueshifted jet versus accretion luminosity. The dashed line plots a (one-sided) ratio $F_j c/L_{\rm acc}$ of 150

from G98 are plotted in Fig. 6 as open circles. The mean ratio $\dot{M}_j/\dot{M}_{acc}$ is now $\simeq 0.1$.

The right-hand panel of Fig. 6 plots the jet momentum-flux (or "thrust") in the jet, $F_j = \dot{M}_j V_j$, versus accretion luminosity, for the same sample. One finds $F_j \simeq 150 L_{acc}/c$ with the updated accretion rates from G98. Note that the correlation would be completely blurred if plotted versus the total luminosity of the system, $L_{bol} = L_* + L_{acc}$, since the strength of jets is uncorrelated with the stellar luminosity L_*, which dominates L_{bol} in most T Tauri stars.

It is noteworthy that the correlations are affected by a large scatter of about a factor 10 in both $\dot{M}_{acc}$ and $\dot{M}_j$. Part of it probably stems from the irregular variability of T Tauri stars: accretion shock diagnostics vary significantly on timescales of days, while forbidden lines trace $\dot{M}_j$ averaged over at least a year (50 AU at 300 km s^{-1}). Another part of the scatter reflects intrinsic uncertainties in the method to derive $\dot{M}_j$: There have been recent efforts to refine jet mass-loss rates using line ratios and spatially resolved or flux-calibrated images. The assumptions and associated uncertainties are described extensively e.g. in [14]. The inferred values typically span an order of magnitude, depending on the distance from the star and on the method used. This may be seen in Fig. 7 in the case of the RW Aur bipolar jets (see also Fig. 10 in [14] for the DG Tau jet, and [82] for the VLA 1 jet). Hence current uncertainties on jet mass-loss rates remain a factor 10 even in the best studied cases. Comparison with the values of $\dot{M}_j$ derived by HEG95 suggest that the latter also suffer a factor 10 uncertainty, though with no large systematic errors (unlike the $\dot{M}_{acc}$ values).

We conclude that, even though single measurements are uncertain by a factor 10, the *average* ejection efficiencies obtained with the HEG95 jet mass-loss rates and the G98 accretion rates should be relatively reliable. Multiplying by 2 to take into account the redshifted lobe of the microjet (usually occulted by the disk), we have:

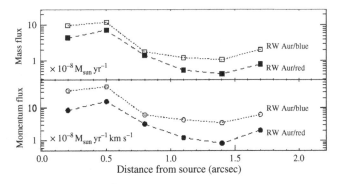

Fig. 7. Mass-loss rate (*top*) and momentum rate (*bottom*) estimates in each of the two jet lobes of RW Aur, as a function of distance from the star. Note the apparent increase by a factor 10 in $\dot{M}_j$ within 0.7 arcsec = 100 AU of the star. From [106]

$$(2\dot{M}_j)/\dot{M}_{acc} \simeq 0.2 \qquad (12)$$

$$(2F_j)c/L_{acc} \simeq 300 \qquad (13)$$

$$(2L_j)/L_{acc} \simeq 0.15 . \qquad (14)$$

$L_j = 1/2\dot{M}_j V_j{}^2$ denotes the kinetic energy flux ("mechanical luminosity") in one jet beam, derived from $F_j c/L_{acc}$ assuming $V_j \simeq 300$km s^{-1}. The scatter in the correlations suggests an uncertainty on these mean values of about a factor 3.

Similar efficiencies were found by Hartigan, Morse, & Raymond (1994) for 3 atomic jets from infrared Class I sources, assuming (i) accretion is strong enough in these embedded objects that it dominates over the stellar photosphere (i.e. $L_{acc} \simeq L_{bol}$), and (ii) Class I sources are close to their final T Tauri mass $\simeq 0.8 M_\odot$ (as confirmed by near-IR spectroscopy of a large Class I sample; cf. [37]). $\dot{M}_{acc}$ is then given by:

$$\dot{M}_{acc} = 1.2 \times 10^{-7} \left(\frac{R_\star}{3R_\odot} \right) \left(\frac{M_\star}{0.8M_\odot} \right)^{-1} \left(\frac{L_{acc}}{L_{bol}} \right) \left(\frac{L_{bol}}{L_\odot} \right) M_\odot \text{yr}^{-1} . \qquad (15)$$

These 3 Class I jets are plotted as starred symbols in Fig. 6, with $\dot{M}_j$ recalculated as in HEG95 for consistency [14]. They extend to higher accretion rates the trend defined by Class II jets in the $\dot{M}_j$-$\dot{M}_{acc}$ and F_j-L_{acc} correlations, suggesting a common ejection mechanism.

3.2 Correlation of Outflow Energetics with L$_{bol}$ in Class 0 Sources

In highly embedded Class 0 protostars , which are in the early vigorous phase of infall, we also expect L$_{bol}$ to be dominated by accretion, at least up to $1000L_\odot$ (see e.g. [83]). On the other hand, evaluating $\dot{M}_{acc}$ is not as straightforward as in the Class I phase: the stellar mass has probably not yet reached its final value, and no near-IR photospheric spectra are available to constrain it. We will thus examine correlations of outflow signatures with L$_{bol}$ only, taken as a tracer of L$_{acc}$.

Because the jet beam is usually heavily extincted or undetected in atomic lines, $\dot{M}_j$ cannot be derived as accurately as in Class I/II jets. Molecular jets (in H$_2$ or CO) may severely overestimate $\dot{M}_j$ if they include entrained/shocked ambient gas. More indirect measures of wind energetics in Class 0 sources have thus been used. They all obey a correlation with L$_{bol}$, hence probably with L$_{acc}$.

We will first discuss the most widely used indicator, namely observations of the bipolar molecular outflows detected in low-excitation CO lines around all Class 0 protostars. A prototypical example is shown in Fig. 8.

The low-velocity (V$_{CO} \simeq 10$ km s^{-1}) and large mass (M$_{CO}$ of up to several $M_\odot$) displayed by CO outflows imply that they do trace material ejected from the source, but rather ambient gas accelerated and swept-up by a collimated jet or wind (cf. [5, 68]). Several facts suggest that jet acceleration

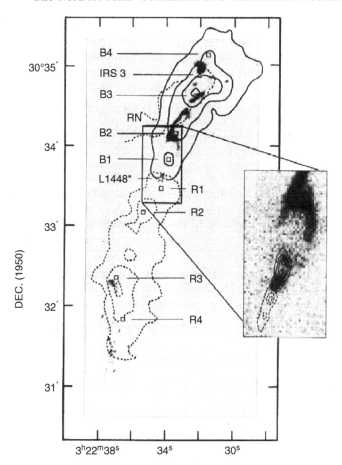

Fig. 8. The bipolar molecular outflow in L 1448 and its associated jet. Low-velocity CO(2-1) emission in the outflow lobes is shown by contours (*solid = blueshifted, dotted = redshifted*), while high-velocity CO "bullets" are denoted by squares. H_2 2.12μm emission from shocked gas is displayed as a greyscale image. The small-scale bipolar SiO jet emanating from the Class 0 source is shown in the insert. Note how molecular gas is shocked and accelerated all along the jet path. From [6]

is important, and perhaps dominant in Class 0 outflows [17, 70], although a "wide-angle" component also appears required to broaden CO outflows to their observed widths in the Class I phase and in luminous sources ([3] and references therein). It has also been proposed that the massive CO outflows in bright sources include a global magnetized recirculation of infalling material (Lery, this volume).

In the following we will assume that CO outflows in Class 0 sources are entirely accelerated by a bipolar jet, keeping in mind that our assumption will overestimate the jet thrust if there is also significant momentum injection by a wide-angle wind unrelated to the jet, or by a circulating flow.

The momentum flux F_{CO} and mechanical luminosity L_{CO} in molecular outflows are correlated with L_{bol} over 5 orders of magnitude [68]. The correlation is tighter when the sample is restricted to well-collimated flows from Class 0 sources [15], where the following least-square fits are obtained:

$$F_{CO} \simeq 4 \times 10^{-5} \left(L_{bol}/L_\odot\right)^{0.7} M_\odot\,\mathrm{km\ s}^{-1}\mathrm{yr}^{-1} \tag{16}$$

$$L_{CO} \simeq 0.04 \left(L_{bol}/L_\odot\right)^{0.8} L_\odot . \tag{17}$$

Updated plots using the enlarged high-luminosity sample in Richer et al. (2000) are shown in Fig. 9. These correlations have been confirmed and extended over a sample of 390 outflows by Wu et al. (2004).

F_{CO} and L_{CO} may be related to the integrated energetics of the driving wind/jets through a model of the wind/cloud interaction, involving both a forward and a reverse shock (see e.g. [62]). In the highly simplified case of highly radiative and planar shocks, ram-pressure equilibrium yields $F_w \simeq F_{CO}$ when $V_{CO} \ll V_j$, while the total mechanical luminosity $L_w \geq 7L_{CO}$ (see e.g. [14]). We thus obtain

$$F_w c/L_{bol} \simeq 1000 \left(L_{bol}/10L_\odot\right)^{-0.3} \tag{18}$$

$$L_w/L_{bol} \geq 0.17 \left(L_{bol}/10L_\odot\right)^{-0.2} . \tag{19}$$

The assumption of highly radiative shocks is supported by the copious emission of outflows from Class 0 sources in far-infrared lines of H_2, CO, and

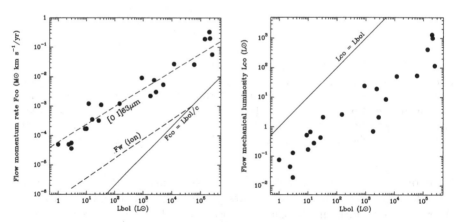

Fig. 9. Left panel: Correlation with the source bolometric luminosity L_{bol} of the momentum rate in the swept-up molecular flow, F_{CO}. For comparison, the top dashed line plots the wind thrust F_w corresponding to the [O I]63μm-derived mass-loss rates ([71], see text) assuming $V_w = 150$ km s^{-1}. The bottom dashed line shows the momentum flux in the ionized part of the jet [84]. **Right panel:** Correlation with L_{bol} of the mechanical luminosity L_{CO} in the swept-up molecular flow. Adapted from [93]

H_2O, with a total luminosity close to L_{CO} [52]. On the other hand the planar approximation is very idealized (see e.g. the curved morphology of H_2 shocks in L 1448, in Fig. 8) hence the above relations are only approximate.

Another indirect diagnostic of wind dynamics that has been used is the outflow luminosity in the [O I]63μm line, expected to be proportional to the mass-flux entering dissociative shocks [63]. Assuming emission from a single shock, current measurements suggest a correlation between $\dot{M}_w$ and L_{bol} (see [21, 71]):

$$\dot{M}_w(63\mu m) \simeq 4 \times 10^{-7}(L_{bol}/L_\odot)^{0.6} M_\odot yr^{-1} . \qquad (20)$$

With a speed of 150 km s^{-1}, the total thrust $\dot{M}_w V_j$ appears compatible with the $F_{CO}-L_{bol}$ correlation (see upper dashed line in the left panel of Fig. 9). Spatially resolved [O I]63μm maps will be very useful to verify these estimates.

The thrust in the ionized part of the jet (derived from the free-free continuum flux) follows a similar correlation with L_{bol}, but shifted downward by a factor 30 ([84]; lower dashed line in the left panel of Fig. 9). Unfortunately, the jet ionization fraction is unknown. Assuming that it is a few %, as found in Class II jets from optical line ratios (e.g. [4]) would give similar values of F_w as above.

3.3 Ejection to Accretion Ratio as a Function of Source Mass and Evolutionary Stage

Two interesting trends may be noted in the momentum efficiency $F_w c/L_{bol}$ of Class 0 sources: (1) a gradual decrease with increasing luminosity of the source, by a factor 10 from $10L_\odot$ to $10^4 L_\odot$ (see Fig. 9); (2) an increase compared with later evolutionary phases, with $F_w c/L_{bol} \simeq 1000$ in Class 0 sources of $L_{bol}= 10L_\odot$, compared with $(2F_j)c/L_{acc} \simeq 300$ in Class I/II jets with similar L_{acc} (cf. Sect. 3.1).

Before interpreting these trends as variations in the accretion-ejection ratio, it is important to realise that the mass of the driving source has an indirect effect on the value of the momentum efficiency. To see this, it is useful to rewrite $F_w c/L_{acc}$ as

$$F_w c/L_{acc} = 1000 \left(\frac{\dot{M}_w}{\dot{M}_{acc}}\right)\left(\frac{V_j}{V_{K,*}}\right)\left(\frac{300km\ s^{-1}}{V_{K,*}}\right) , \qquad (21)$$

where $V_{K,*} = \sqrt{GM_*/R_*}$ is the keplerian velocity at the stellar surface. Even if the ejection-accretion properties (which roughly determine $\dot{M}_w/\dot{M}_{acc} \times(V_j/V_{K,*})$) remain constant, we see that the observed trends could be obtained if $V_{K,*}$ *increases* overall with the luminosity of the source, and if it *decreases* in low-luminosity Class 0 sources compared with Class I's. Both hypotheses appear quite reasonable if we consider that (i) the brightest Class 0 sources are probably massive protostars of up to $\geq 10M_\odot$, (ii) the massive envelopes of Class 0 sources indicate an early stage of infall where the final stellar mass is not yet assembled, in contrast to Class I sources [1]; hence one does expect a significantly smaller M_* (and $V_{K,*}$) in Class 0's on average. At a given L_{acc}, Class 0 sources would then have a higher accretion rate than Class I's.

Another factor possibly contributing to the lower momentum efficiency at high L_{bol} is that the photospheric luminosity L_* may exceed L_{acc} in massive protostars (which reach the main-sequence while still accreting; [83]). This will diminish $F_w c/L_{bol}$ compared to the actual momentum efficiency with respect to accretion, $F_w c/L_{acc}$, given by (21).

Simple calculations show that these two effects, combined, could explain the magnitude of the decrease in $F_w c/L_{bol}$ with evolutionary phase and with L_{bol}, without having to invoke a major change in ejection/accretion properties [12, 61, 93]. Bontemps et al. (1996) found a tight correlation between F_{CO} and M_{env} across Class 0 and Class I sources, and inferred $\dot{M}_w/\dot{M}_{acc} \simeq 0.1 \left(\tau_{env}/10^5 \mathrm{yr}\right) \left(V_j/150 \mathrm{kms}^{-1}\right)^{-1}$, where $\tau_{env} = M_{env}/\dot{M}_{acc}$ is the characteristic decay time of the envelope, assumed constant (a more detailed model is presented in Henriksen et al. 1997). Using a different approach, Richer et al. (2000) estimated that $(\dot{M}_w/\dot{M}_{acc})(V_j/V_{K,*}) \simeq 0.3$ in Class 0 outflow sources across the whole range of L_{bol}.

Given the assumptions involved, this is not a definite proof that changes in ejection/accretion properties do not occur, but it raises the theoretically appealing possibility that a single mechanism could be responsible for the jets observed across the whole mass spectrum and evolutionary sequence of YSOs.

3.4 Summary

- Jet energetics in Class II sources are clearly correlated with accretion. The mean mass, momentum, and energy efficiencies are (within a factor 3): $(2\dot{M}_j)/\dot{M}_{acc} \simeq 0.2$, $(2F_j)c/L_{acc} \simeq 300$, $(2L_j)/L_{acc} \simeq 0.15$. These values pertain only to the optically bright part of the jet.
- Jets in low-luminosity Class I sources appear to have similar accretion efficiencies as jets from Class II sources. The jet speeds are also similar.
- Outflow signatures in deeply embedded Class 0 sources show clear correlations with the source bolometric luminosity over 5–6 orders of magnitude, probably tracing an underlying ejection-accretion correlation. The momentum efficiency is high, $F_w c/L_{bol} \simeq 1000(L_{bol}/10L_\odot)^{-0.3}$. Reasonable assumptions about the stellar mass or age suggest $(\dot{M}_w/\dot{M}_{acc})(V_j/V_{K,*}) \simeq 0.1$–$0.3$ in Class 0 sources of all luminosities, similar to Class I/II jets within the uncertainties.
- If momentum injection in CO outflows during the Class 0 phase is dominated by the jets (and not by wide-angle winds or circulation flows), the data appear consistent with a single jet mechanism operating at all masses and all evolutionary stages of YSOs.

4 Comparison with HD and MHD Ejection Models

We will now discuss (in chronological order) the various hydrodynamical and MHD processes that have been proposed for the acceleration of jets from young stars. We will examine the problems that they encounter to reproduce the observed jet speeds and/or large ejection-accretion efficiencies (see previous section). Again we will focus on the case of low-luminosity Class II stars, where the constraints are more severe. It will be concluded that only MHD acceleration processes are readily able to reproduce observations.

4.1 Radiation Pressure

Absorption of stellar photons transfers photon momentum flux to the gas, thus exerting a net outward force on it. Assuming an isotropic radiation source, the net force *per unit volume of gas* at distance R from the source is:

$$f_{rad} = (L_{\text{bol}}/c)/(4\pi R^2 L) = \kappa_R \rho (L_{\text{bol}}/c)/(4\pi R^2) . \tag{22}$$

Here L_{bol}/c is the total photon momentum entering the spherical surface per unit time (each photon carries a momentum $h\nu/c$), and $4\pi R^2 L$ is the volume of gas over which this momentum is absorbed, with $L = 1/\rho\kappa_R$ the photon mean free-path (we denote by κ_R the Rosseland mean gas opacity per gram of matter ($\text{cm}^2 \text{ g}^{-1}$) averaged over the local radiation spectrum).

To drive a wind, the radiation pressure force must overcome the gravity of the central star, i.e. $f_{rad} > \rho GM_\star/R^2$. This condition gives the "Eddington" luminosity, above which radiation pressure alone will expel circumstellar matter:

$$L_{Edd} = 4\pi cGM_\star/\kappa_R = 1200 L_\odot (M_\star/M_\odot)(10\text{cm}^2 \text{ g}^{-1}/\kappa_R) , \tag{23}$$

where the adopted numerical value for κ_R is typical for ISM dust grains near their sublimation temperature. The above condition clearly fails for low-luminosity Class I and Class II jet sources, where $M_\star \simeq M_\odot$ and $L_{\text{bol}} < 100\ L_\odot$. It could be fulfilled, however, in objects with very large accretion rates: Assuming that L_{bol} is dominated by the accretion luminosity $L_{\text{acc}} = GM_\star \dot{M}_{\text{acc}}/R_\star$, the Eddington luminosity is exceeded if

$$\dot{M}_{acc} > 4\pi cR_\star/\kappa_R = 1.2 \times 10^{-4} M_\odot \text{yr}^{-1}(R_\star/3R_\odot)(10\text{cm}^2 \text{ g}^{-1}/\kappa_R) . \tag{24}$$

Thus, radiation pressure could significantly counteract gravity near young and/or massive protostars accreting at very high rates.

A more fundamental problem with radiation pressure, which was noted early on [68], is that it fails to explain the very large momentum transfer efficiency observed in jets and CO outflows, where $F_w c/L_{\text{bol}}$ typically reaches $100 - 1000$. Integrating the expression for f_{rad} in (22) over the entire wind volume one finds

$$F_{rad} = (L_{bol}/c) \times \int \rho(R)\kappa_R(R)dR = (L_{bol}/c) <\tau> , \qquad (25)$$

where $<\tau>$ is the effective opacity of the whole wind to radiation.

The UV-optical photons emitted by the star + accretion shock are initially absorbed efficiently by the dust grains that they first encounter, i.e. the value of κ_R is highest at the dust sublimation radius. However, the absorbed energy is then re-radiated by the grains as a blackbody at their own equilibrium temperature, i.e. mainly in the infrared. Dust absorption is less and less efficient at longer wavelengths (it varies roughly as λ^{-1}) so κ_R decreases steeply as one moves away from the source. As a result, the effective opacity of the wind to radiation, $<\tau>$, does exceed 1–10 (e.g. [81]). This is insufficient compared to observations, except perhaps for the most luminous protostars with $L_{bol} > 5 \times 10^4 \, L_\odot$, where CO outflows have $F_{CO} \simeq 10 \, L_{bol}/c$ (see [93]). One may thus safely rule out radiation pressure as the generic driving agent of YSO jets.

4.2 Thermal Pressure Gradients

A second HD acceleration mechanism relies on thermal pressure gradients at the base of the wind: If the sound speed at the wind base is of order of the escape speed, steady solutions exist where the gas is accelerated through a sonic point and escapes to infinity with non-zero velocity. A classic example is the Parker solution for the solar wind discussed in K. Tsinganos's lecture (this volume).

The temperature at the wind base required to reach a given asymptotic speed V_j may be evaluated through energy conservation along a streamline (Bernoulli's invariant). Neglecting initial rotation with respect to gravity (e.g. T Tauri stars rotate well below break-up), we have

$$\frac{V_j^2}{2} + H = H_0 - GM_\star/R_\star , \qquad (26)$$

where H_0 and H are the specific enthalpy at the wind base and at infinity. Following Ferreira, Dougados, and Cabrit (2006) we may calculate the corresponding temperature by introducing a "heating parameter" β, measuring the ratio of enthalpy given to the flow to initial gravitational energy

$$\beta = 2(H_0 - H)/(GM_\star/R_\star) . \qquad (27)$$

The asymptotic wind speed then writes

$$V_j = \sqrt{\beta - 2}\sqrt{GM_\star/R_\star} \simeq 250 \sqrt{\beta - 2} \left(\frac{M_\star}{M_\odot}\right)^{1/2} \left(\frac{R_\star}{3R_\odot}\right)^{-1/2} \text{ km s}^{-1} . \qquad (28)$$

Observed speeds $\simeq 300$ km s^{-1} thus require $\beta \simeq 3$, and we obtain the minimum initial temperature T_0 (assuming $H \ll H_0$):

$$T_o = 1.5 \times 10^6 \left(\frac{\beta}{2}\right) \left(\frac{M_\star}{M_\odot}\right) \left(\frac{R_\star}{3R_\odot}\right)^{-1} \text{K} . \qquad (29)$$

Such high "coronal" temperatures may well be present at the surface of convective young stars, either as a result of solar-type magnetic surface activity, or immediately behind the accretion shock (before significant cooling occurs). And indeed, there is recent evidence for accretion-related hot coronal winds in T Tauri stars, in the form of of P Cygni profiles or blueshifted deficits in Helium lines and UV lines [39, 42].

However, as first pointed out in a seminal paper by De Campli (1981), one faces a severe problem to accelerate *all* of the jet mass-loss rate by this mechanism: The combination of high coronal temperatures and high jet densities should produce extremely strong X-ray Bremsstrahlung emission, which is not observed. Spherical models of thermally-driven coronal winds in T Tauri stars from Bisnovatyi-Kogan and Lamzin (1977) predict typical X-ray luminosities of (cf. Table 4 of [34]):

$$L_{\text{corona}} \simeq 5 \times 10^{34} \text{ erg s}^{-1} \left(\frac{\dot{M}_w}{2 \times 10^{-8} M_\odot \text{yr}^{-1}}\right)^2 . \qquad (30)$$

where the $\dot{M}_w{}^2$ dependence stems from the n_e^2 dependence of optically thin free-free radiation. Observed X-ray luminosities from T Tauri stars are $L_X < 10^{31}$ erg s^{-1}, and generally 10 times smaller [43, 64]. Hence thermally driven coronal winds cannot have mass-loss rates above a few 10^{-10} $M_\odot \text{yr}^{-1}$. We have seen that typical jet mass-loss rates in T Tauri stars exceed this value by 2 orders of magnitude.

The discrepancy between predicted and observed X-ray emission will worsen for sources more luminous than T Tauri stars. Observations indicate that $\dot{M}_w$ scales as $L_{\text{bol}}^{0.6}$ (see (20)). Hence the predicted L_{corona} increases as $L_{\text{bol}}^{1.2}$, while the observed X-ray flux increases on average only as $L_X \propto L_{\text{bol}}$ [43].

The problem of excessive X-ray flux may be avoided in alternative models where the wind is "cold" at its base (thus producing negligible X-rays), and enthalpy is provided to the flow further up (e.g. [95]). As discussed by Ferreira et al. (2006), the same equations as above still apply, provided the β term also includes the specific heat input along the streamline, denoted $\mathcal{F}(s)$, i.e. $\beta = 2(\mathcal{F}(s) + H_0 - H)/(GM_\star/R_\star)$. However, this type of model meets a more fundamental limit of pressure driven stellar winds, namely their *low efficiency*.

The net "heating" power transferred to the two jet beams, each with one-sided mass-flux $\dot{M}_j$, may be written

$$L_\beta = \frac{\beta}{2}\frac{GM_\star}{R_\star}2\dot{M}_j = \frac{\beta}{\beta-2}V_j^2\dot{M}_j = \frac{\beta}{\beta-2}(2L_j) , \qquad (31)$$

where we have used (28) for V_j, and $2L_j$ is the mechanical power in both jets. Since $\beta \simeq 3$, it appears that 3 times more energy must be injected in heat than will eventually end up in kinetic energy ... The net heating power L_β may also be expressed in terms of the accretion luminosity $L_{\rm acc}$ [47] as

$$L_\beta = \frac{\beta}{2}\frac{GM_\star}{R_\star}2\dot{M}_j = \beta\left(\frac{\dot{M}_j}{\dot{M}_a}\right)L_{\rm acc} . \tag{32}$$

With $\beta \simeq 3$ and one-sided $\dot{M}_j/\dot{M}_{\rm acc}\simeq 0.1$, the net transferred power would represent 30% of the accretion luminosity. As no heating process is "loss-free" (e.g. radiative losses must be compensated for), the actual power required to accelerate the jets should be several times larger, becoming comparable to $L_{\rm acc}$. It is very difficult to envision a heating process efficient enough to deposit a major fraction of the accretion energy at high altitudes in the wind (in the next section, we will see that the problem remains if pressure is provided by MHD waves).

For completeness, it is interesting to mention another site where thermally driven (but slow) winds may occur in YSOs: that is when a hot ionized layer is produced at the disk surface through irradiation by a strong UV field. Thermal balance in the photo-ionized layer establishes a temperature of order 10^4 K, corresponding to a sound speed of $c_s \simeq 10$ km s^{-1}, so no excess X-ray emission is predicted. Material is lifted off the disk beyond the critical radius $r_g \simeq 10$ AU where the sound speed equals the keplerian speed (see [48] and references therein). Since terminal velocities are $\simeq 3C_s = 30$ km s^{-1}, this "disk evaporation" mechanism seems promising to explain the low-velocity component of forbidden lines in T Tauri stars. However, it is of course unable to provide the high speed material at several 100 km s^{-1} in jets. The mass-loss rate is also very small ($10^{-10}M_\odot{\rm yr}^{-1}$ for typical T Tauri parameters; [48]).

Therefore, while thermally-driven winds from a hot stellar corona and from the irradiated disk surface may well be present in YSOs (as it does appear to be the case in T Tauri stars), they do not seem capable of explaining the large ejection/accretion ratio in the high-speed jets.

4.3 Alfvén-wave Pressure Gradients

To avoid the excessive radiative X-ray losses associated with thermally driven winds, De Campli (1981) suggested an alternative mechanism involving the deposition of momentum of action-conserving MHD waves into the mean flow, providing an anisotropic effective pressure gradient which accelerates the flow.

Alfvén waves appear as the most promising agent, as they are not compressible and thus have large damping lengths, allowing them to deposit momentum and energy into the wind before dissipating. Since young stars possess both deep convective layers and rather strong magnetic fields (up to 1 kG at the footpoint of accretion columns), the excitation of MHD waves at the stellar surface or in the magnetosphere indeed appears plausible.

De Campli calculated steady, spherical, super-Alfvénic wind solutions and found that wind acceleration up to terminal velocities $\simeq 300$ km s^{-1} could be achieved if (i) the Alfvén waves are coherent and in phase over a large solid angle (in order not to diverge too fast), (ii) the initial B-field intensity is a few 100 G (in order to avoid non-linear wave dissipation before the wind becomes super Alfvénic). Under such conditions, he obtained that the efficiency of energy transfer from the wave flux to the wind was typically 10%–20%, for B_0 of 150–500 G. In other words,

$$L_{\text{wave}} \simeq 5 - 10L_w .\qquad(33)$$

We have seen that the total jet mechanical luminosity is typically 0.1 of L_{acc}. Therefore accelerating the jets by this process would require a luminosity in *coherent* Alfvén waves L_{wave} of 0.5–1 L_{acc}. This sounds uncomfortably large, given that incoherent Alfvén waves and dissipative waves (acoustic, magnetosonic...) will necessarily be excited as well, inducing additional losses. Therefore, acceleration by Alfvén-wave pressure is not an efficient driving mechanism for YSO jets.

Note that one is basically faced with a similar efficiency problem as in winds driven by thermal pressure gradients. This is not too surprising, as the effect of Alfvén wave pressure gradients acts in the Bernoulli conservation equation in a similar way as thermal pressure gradients, simply adding to the heating term β an extra term describing the transfer of energy from the Alfvén waves to the flow (see (38) in [34, 47]).

4.4 Magneto-centrifugal MHD Acceleration

To overcome the efficiency problem that plagues pressure-driven winds, a radically different acceleration mechanism has been envisioned, involving magnetic torques. A strong large-scale field is anchored in a rotating object (star, disk, infalling envelope). The magnetic field exerts a braking torque, removing angular momentum from the rotating object, and transferring angular momentum to the outflowing gas. The centrifugal and Lorentz forces then accelerate the flow along the field lines, making it eventually become super-Alfénic [107].

This so-called "magneto-centrifugal" acceleration process described in greater detail in the various theoretical lectures in this volume, is "loss-free" in the sense that all (or most) of the rotation energy extracted from the source is eventually converted into wind kinetic energy.

Below I confront in more detail the jet observations with model predictions for MHD acceleration along field lines anchored (1) in the disk, or (2) in the star. Models where the field lines are anchored in the infalling envelope are discussed in the lecture by Lery (this volume).

1- Field Lines Anchored in the Disk

Magneto-centrifugal ejection from accretion disks was first proposed as the source of YSO jets and outflows by Pudritz and Norman (1983). This hypothesis has been studied in great detail since then (see [87] for a comprehensive review). One of its most attractive aspects is the possibility of a high ratio of jet to accretion power. If wind magnetic torques extract most of the angular momentum from the accretion flow, the jet mechanical power will be essentially the rate of energy removal between the outer and inner radii of the launch region (see e.g. [45]):

$$2L_j \simeq -\frac{GM_\star \dot{M}_a(r_{out})}{2r_{out}} + \frac{GM_\star \dot{M}_a(r_{in})}{2r_{in}} \simeq \frac{GM_\star \dot{M}_a(r_{in})}{2r_{in}} \tag{34}$$

where the $1/2$ factor derives from the fact that the total (gravitational + kinetic) specific energy is $-GM_\star/2r$ for a keplerian flow. Thus, one can easily reach the high observed efficiencies $2L_j \simeq 0.1$ L_{acc} if the inner edge of the wind launch region is at $r_{in} \simeq 5$ $R_\star$.

The final jet speed depends on the "magnetic lever arm parameter" $\lambda \simeq (r_A/r_0)^2$, which measures the ratio of extracted vs. initial specific angular momentum (here r_0 is the launch radius and r_A the (cylindrical) Alfvén radius). When all of the Poynting flux has been transferred to the matter, the final jet speed is given by [11]:

$$V_j = \sqrt{GM_\star/r_0}\sqrt{2\lambda - 3} \ . \tag{35}$$

The typical observed maximum jet speeds $\simeq 300$ km s^{-1} would then indicate $\lambda \simeq 5$ for $r_{in} \simeq 5$ $R_\star$ in a typical low-mass YSO. Such a moderate value of λ is compatible with current estimates/upper limits on rotation speeds in jets (see [47] and references therein).

Self-similar models including the keplerian disk vertical equilibrium show that the value of λ is inversely related to the efficiency of mass-loading onto the field lines through $\lambda \simeq 1 + 1/2\xi$, where $\xi = \text{dlog}(2\dot{M}_j)/\text{dlog}r$, as long as $\xi \ll 1$ [44]. Hence observed jet speeds would suggest a rather high mass-load with $\xi \simeq 0.1$. The total ejection to accretion ratio is then

$$2\dot{M}_j/\dot{M}_{acc} = \xi \ln(r_{out}/r_{in}) \ , \tag{36}$$

compatible with observations if $r_{out}/r_{in} \leq 10$. Note that this constraint applies only to the launch region producing the *observable optical* jet. Material launched from radii beyond 1 AU would escape detection if it is molecular, or too cold to emit.

A high mass-load with $\xi \simeq 0.1$ can only be achieved with a source of heating near the disk surface (Casse & Ferreira 2000; see also Ferreira's lecture, this volume). The required heating only represents a tiny fraction of the accretion power, so it is energetically possible, but its nature remains to be clarified (possible sources include X-rays from the star, and dissipation of MHD turbulence).

A possible variant proposed by Shu et al. (1994) is that the centrifugal wind is launched not from an extended disk region, but from a narrow annulus at the inner edge where the disk is truncated by the stellar magnetosphere.

The flow is highly non self-similar, hence the Alfvén surface is obtained from a prescribed mass-loading function. The asymptotic speeds would be $\simeq 200$–350 km s^{-1} ($\lambda \simeq 3-6$) for a ratio $2\dot{M}_j/\dot{M}_{\mathrm{acc}} = 0.3$ [96], also compatible with observations. The main observable difference with winds launched over a large range in disk radii is that it predicts higher poloidal speeds and smaller rotation signatures in the outer regions of the jet [47, 89, 97].

2- Field Lines Anchored in the Star

Hartmann & MacGregor (1982) proposed that energetic flows from young stars could be powered by magneto-centrifugal ejection from the surface of magnetized protostars rotating close to break-up. Concentrating on very massive protostars ($M_\star = 10 M_\odot$ and $L_{\mathrm{bol}} \simeq 5 \times 10^4 L_\odot$) they obtained wind speeds of a few 100 km s^{-1} for a 10 G field, and outflow momenta compatible with observations for a rotation rate at $\geq 90\%$ of break-up. The attractive aspect is that it provides a natural means to spin-down protostars in less than 10^5yrs, as required by the slow rotation rates of Class I sources [28]. Matt & Pudritz (2005) recently proposed that strong magnetized stellar winds from T Tauri stars could similarly prevent them from spinning-up by accretion.

One still faces two problems in trying to explain all YSO jets in this picture: (1) the correlation with accretion would have to be somewhat indirect, as the mass-loss rate is governed by the density at the slow point near the stellar surface, not by the density of any accretion flow, (2) the efficiency of the mechanism depends critically on the stellar rotation rate. For example, Hartmann & MacGregor (1982) find that the mass-loss rate and the wind thrust drop drastically (by, respectively, 6 and 4 orders of magnitude) when the rotation rate decreases only slightly, from 0.9 to 0.75 of break-up, all other parameters being held constant. Increasing the B-field increases the terminal speed, but has no effect on the mass-loss rate. This would not easily explain why jets from Class I/II stars have similar momentum efficiencies (within a factor 3–10) as outflows from Class 0s, unless the Class 0 also rotate well below break-up — in which case the problem of initial protostellar spin-down remains unsolved. Compensating the slower rotation by changes in stellar parameters (e.g. density and temperature at the slow point) would require very fine tuning.

An alternative model possibly solving both problems is the Reconnexion X-wind ("ReX wind") explored by Ferreira et al. (2000). It assumes that matter is loaded onto stellar field lines not at the stellar surface, but very near the *corotation point* in the disk where the keplerian angular velocity matches the stellar rotation frequency. The (sporadic) mass-loading would occur by magnetic reconnexion between closed stellar field lines and open disk field, at an equatorial magnetic X-point in the disk (see Fig. 9 and Sect. 4.6 in Ferreira, this volume). As in the Hartmann & MacGregor case, this wind is ultimately powered by the spin-down of the star, not by accretion energy. However, mass-loading in the "ReX wind" is now occuring in the disk, hence some correlation of the ejected mass-flux $\dot{M}_X$ with the accretion rate might be expected. Furthermore, the corotation point is by definition always rotating at break-up, so the mass-loss efficiency may not vary excessively.

Assuming $2\dot{M}_X/\dot{M}_{\mathrm{acc}} \simeq 0.1$, a magnetic lever arm parameter $\lambda \simeq 3$, and an initial accretion rate of $10^{-5} M_\odot \mathrm{yr}^{-1}$, Ferreira (2000) finds that a low-mass protostar is spun-down to 20%–10% of break-up in $10^5 - 10^6$ yrs, as observed, while the disk accretion rate drops to 10^{-7}–10^{-8} $M_\odot \mathrm{yr}^{-1}$, consistent with the mean behavior of $\dot{M}_{\mathrm{acc}}$ in T Tauri stars [60]. The terminal speeds of 150–300 km s^{-1} agree with jet observations. In this model, the open disk field is assumed to be constantly replenished by an MHD disk wind operating beyond corotation, so one should in fact observe both contributions (sporadic ReX wind and steady disk wind) in the jets.

4.5 Relaxation of Twisted Magnetospheric Field ("Magnetic Tower")

The last category of ejection mechanisms proposed in YSOs relies on the interaction of closed loops of magnetospheric stellar field with the inner disk near the magnetopause, in the configuration where the stellar magnetic moment is *anti-parallel* to the disk field, or when there is no significant disk field (see Fig. 9 in Ferreira, this volume).

In this geometry, any small differential "twist" between the star and the inner disk edge triggers a huge stretching of the magnetic loops at an angle of 60° from the polar axis [73]. The complex time-dependent evolution of this interaction has been the subject of intensive numerical simulations, nicely reviewed in Pudritz et al. (2006). The general conclusion is that the resulting outflows are essentially uncollimated, hence the formation of a tall jet ("magnetic tower") by this process would require external confinement (e.g. [65, 74]). We have seen this is problematic around Class II sources. Axial narrow "jets" are seen in some simulations, but they appear to be an intermittent feature. Recent laboratory experiments also suggest that the polar "jet" is very transient and susceptible to kink instabilities [69].

Simulations by Matt et al. (2002) do not reveal a clear correlation between the disk density and the mass-flux periodically injected inside the "magnetic tower" by magnetic reconnexion, unlike what would be needed to reproduce the accretion-ejection correlation. However, as the twisted loop stretches out, it provides open field lines to the inner disk, possibly allowing the launching of an accretion-powered magneto-centrifugal disk wind there [98]. The properties of such an MHD disk wind were described above in Sect. 4.4.

4.6 Summary

- Acceleration by radiative pressure on dust is excluded, as it cannot provide the high momentum efficiencies $F_w c/\mathrm{L_{bol}} > 10$ observed in YSO jets.

- Hydrodynamic acceleration of the jets by thermal or wave pressure gradients requires an energy input rate in heat or coherent Alfvén waves that is at least comparable to L_{acc}. This sounds difficult to achieve given that some additional losses (radiative cooling, excitation of dissipating waves) will necessarily occur. More tenuous ejections from the stellar corona or the disk surface are however possible (and likely).

- MHD magneto-centrifugal acceleration offers the most efficient acceleration process. An MHD disk wind launched around $5R_*$ and beyond most readily explains the high ejection-accretion efficiencies $2L_j/L_{acc} \geq 0.1$ at all phases of evolution, as well as observed jet speeds. Centrifugal acceleration along open stellar field-lines could also contribute significantly to the observed jets, if it is the main agent responsible for braking down young stars. The resulting MHD jets will be self-collimated.

- MHD relaxation of twisted magnetospheric loops threading the inner disk edge may result in plasmoid ejection at mid-latitudes, but the correlation with accretion rate is unclear, and the outflow is not intrinsically collimated.

5 General Conclusions

We have found that hydrodynamic pressure cannot explain the collimation properties of jets in sources with and without infalling envelopes. Hydrodynamic acceleration (by pressure gradients) also does not appear efficient enough to reproduce the high ratios of jet momentum and jet power to accretion luminosity that are observed at all phases of evolution of YSOs. Therefore, MHD processes appear definitely needed to collimate and accelerate YSO jets.

Among possible MHD processes, magneto-centrifugal driving appears as the most promising, as it provides both acceleration and self-collimation compatible with observations. The observed accretion-ejection efficiencies suggest that launching occurs from the inner disk regions, where it would be directly powered by the release of accretion energy. Such a process could be universal to young accreting stars of all masses and ages, and could start very early on after the formation of a hydrostatic stellar core [7, 76, 87, 104].

The fact that young stars do not spin up despite continuing accretion has lead to suggestions that a centrifugal MHD wind along open stellar field lines might also be present, perhaps as a (reconnexion-fed) inner flow within the disk wind. Simple calculations indicate that the contribution to observed jet mass-fluxes could be substantial, although further modelling of the mass loading and dynamics is critically needed.

MHD relaxation of twisted magnetospheric loops does not appear to produce by itself the launching of a jet flow comparable to observations, but it could provide the open field lines required for centrifugal acceleration.

Therefore, the ubiquitous existence of YSO jets appears intimately related with magnetic extraction of angular momentum from the inner disk around (and possibly the surface of) accreting young stars, from the earliest stages of star formation all the way through the pre-main sequence classical T Tauri phase.

References

1. André, P., Ward-Thompson, D., Barsony, M. 2000: From prestellar cores to protostars: The initial conditions of star formation. In *Protostars & Planets IV*, ed. by V. Mannings, A.P. Boss, S.S. Russell, (University of Arizona Press, Tucson), pp. 59–96
2. Anglada, G. 1996: Radio Jets in Young Stellar Objects. In *Radio emission from the stars and the sun*, ed. by A. R. Taylor and J. M. Paredes, ASP Conf Series, Vol. 93, p. 3
3. Arce, H.G. et al. 2006: Molecular outflows from low- to high-mass star formation. In *Protostars & Planets V*, ed. by B. Reipurth, D. Jewitt, K. Keil (University of Arizona Press, Tucson), pp. 245–260
4. Bacciotti, F., Eisloeffel, J. 1999, A&A 342, 717
5. Bachiller, R., Tafalla, M. 1999: Bipolar Molecular Outflows. In *The Origin of Stars and Planetary Systems*, ed. by Charles J. Lada and Nikolaos D. Kylafis (Kluwer Academic Publishers), p. 227
6. Bally, J., Lada, E.A., Lane, A.P. 1993, ApJ 418, 322
7. Banerjee, R., Pudritz, R. 2006, ApJ 641, 949
8. Barral, J.F., Cantó, J. 1981, RMAA 5, 101
9. Basu, C. & Mouschovias, T.C. 1994, ApJ 432, 720
10. Bisnovatyi-Kogan, G.S., Lamzin, S.A. 1977, Astr.Zh. 54, 1268
11. Blandford, R.D., Payne D.G. 1982, MNRAS 199, 883
12. Bontemps, S., Andre, P., Terebey, S., Cabrit, S. 1996, A&A 311, 858
13. Burrows, C.J. et al. 1996, ApJ 473, 437
14. Cabrit, S. 2002: Constraints on Accretion-Ejection structures in young stars. In: *Star Formation and the Physics of Young Stars*, ed. by J. Bouvier, J.-P. Zahn (EDP Sciences, Les Ulis 2002) pp. 147–182
15. Cabrit, S., Bertout, C. 1992, A&A, 261, 274
16. Cabrit, S., Ferreira, J., Raga, A.C. 1999, A&A 343, L61
17. Cabrit, S., Raga, A., Gueth, F. 1997: Models of bipolar molecular outflows. In *H*erbig-Haro flows and the birth of low mass stars, eds. B. Reipurth and C. Bertout (Kluwer, Dordrecht) p. 163
18. Cabrit, S., Edwards, S., Strom, S.E., Strom, K.M. 1990, ApJ 354, 687
19. Cabrit, S., Pety, J., Pesenti, N., Dougados, C. 2006, A&A, 452, 897
20. Caratti o Garatti, A., Giannini, T., Nisini, B., Lorenzetti, D. 2006, A&A 449, 1077
21. Ceccarelli, C., Haas, M.R., Hollenbach, D.J., Rudolph, A.L. 1997, ApJ 476, 771
22. Chernin, L.M. 1995, ApJ 440, L97
23. Cesaroni, R., Felli, M., Jenness, T., Neri, R., Olmi, L., Robberto, M., Testi, L., Walmsley, C.M. 1999 A&A, 345, 949

24. Chandler, C.J., Richer, J.S. 2001, ApJ 555, 139
25. Claussen, M.J., Marvel, K.B., Wootten, A., Wilking, B.A. 1998, ApJ 507, L79
26. Cohen, M., Emerson, J.P., Beichman, C.A. 1989, ApJ 339, 455
27. Corcoran, M., Ray, T.P. 1998b, A&A, 331, 147
28. Covey, K.R., Greene, T.P., Doppmann, G.W., Lada, C.J. 2005, AJ 129, 2765
29. Dartois, E., Dutrey, A., and Guilloteau, S. 2003, A&A 399, 773
30. Davis, C.J., Ray, T.P., Desroches, L., Aspin, C. 2001, MNRAS 326, 524
31. Davis, C.J., Stern, L., Ray, T.P., Chrysostomou, A. 2002, A&A 382, 1021
32. Davis, C.J., Whelan, E., Ray, T.P., Chrysostomou, A. 2003, A&A 397, 693
33. Davis, C.J., Varricatt, W.P., Todd, S.P., Ramsay Howat, S.K. 2004, A&A 425, 981
34. De Campli, W.M. 1981, ApJ 244, 124
35. Delamarter, F. & Hartmann, L. 2000, ApJ 530, 923
36. Devine, D., Bally, J., Reipurth, B., Heathcote, S. 1997, AJ 114, 2095
37. Doppmann, G.W., Greene, T.P., Covey, K.R., Lada, C.J. 2005, AJ 130, 1145
38. Dougados, C., Cabrit, S., Lavalley-Fouquet, C., Ménard, F. 2000, A&A 357, L61
39. Dupree, A.K., Brickhouse, N.S., Smith, G.H., Strader, J. 2005, ApJ 625, L131
40. Dutrey, A. et al. 1998, A&A 338, L63
41. Dyson, J. 1984, Ap&SS 106, 181
42. Edwards, S. et al. 2003, ApJ 599, L41
43. Feigelson, E.D., Montmerle, T. 1999, ARAA 37, 363
44. Ferreira, J. 1997, A&A 319, 340
45. Ferreira, J., Pelletier, G. 1995, A&A 295, 807
46. Ferreira, J., Pelletier, G., Appl, S. 2000, MNRAS 312, 387
47. Ferreira, J., Dougados, C., Cabrit, S. 2006, A&A, 453, 785
48. Font, A., McCarthy, I.G., Johnstone, D., Ballantyne, D.R. 2004, ApJ 607, 890
49. Furuya, R.S., Kitamura, Y., Saito, M.,Kawabe, R., Wootten, H.A. 1999, ApJ 525, 821
50. Furuya, R.S., Kitamura, Y., Wootten, H.A., Claussen, M.J., Kawabe, R. 2001, ApJ 559, L143
51. Garcia, P., Cabrit, S., Ferreira, J., Binette, L. 2001, A&A 377, 609
52. Giannini, T. et al. 2001, ApJ 555, 40
53. Gueth, F., Guilloteau, S. 1999, A&A 343, 571
54. Guilloteau, S., Bachiller, R., Fuente, A., Lucas, R. 1992, A&A 265, L49
55. Gullbring, E., Hartmann, L., Briceño, C., Calvet, N. 1998, ApJ 492, 323 (G98)
56. Hartigan, P., Edwards, S., 2004, ApJ 609, 261
57. Hartigan, P., Morse, J., Raymond,. J. 1994, ApJ 436, 125
58. Hartigan, P., Edwards, S., Gandhour, L. 1995, ApJ 452, 736 (HEG95)
59. Hartmann, L., MacGregor, K.B. 1982, ApJ 259, 180
60. Hartmann, L., Calvet, N., Gullbring, E., D'Alessio, P. 1998, ApJ 495, 385
61. Henriksen, R.; Andre, P.; Bontemps, S. 1997, A&A 323, 549
62. Hollenbach, D. J. 1997: The physics of molecular shocks in YSO outflows in Herbig-Haro flows and the birth of low mass stars eds. B. Reipurth and C. Bertout (Kluwer, Dordrecht) p. 181
63. Hollenbach, D., McKee, C.F. 1989, ApJ 342, 306
64. Imanishi, K. et al. 2003, PASJ 55, 653
65. Kato, Y., Mineshige, S., Shibata, K. 2004, ApJ 605, 307
66. Kwan, J., Tademaru, E. 1988, ApJ 332, L41

67. Königl, A. 1982, ApJ, 261, 115
68. Lada, C.J. 1985, ARAA, 23, 267
69. Lebedev, S.V. et al. 2005, MNRAS 361, 97
70. Lee, C.-F., Stone, J.M., Ostriker, E.C., Mundy, L.G. 2001, ApJ 557, 429
71. Liseau, R. et al. 1997: Far-IR spectrophotometry of HH flows with the ISO long-wavelength spectrometer. In *Herbig-Haro flows and the birth of low mass stars* eds. B. Reipurth and C. Bertout (Kluwer, Dordrecht) p. 111
72. Lynden-Bell, D., Pringle, J.E. 1974, MNRAS 168, 603
73. Lynden-Bell, D., Boily, C. 1994, MNRAS 267, 146
74. Lynden-Bell, D. 2003, MNRAS 341, 1360
75. Marti, J., Rodriguez, L.F., Reipurth, B. 1998, ApJ 502, 337
76. Matsumoto, T., Tomisaka, K. 2004, ApJ 616, 266
77. Matt, S., Pudritz, R. 2005, ApJ 632, L135
78. Matt, S., Goodson, A.P., Winglee, R.M., Böhm, K.-H. 2000, ApJ 574, 232
79. Ménard, F., Duchêne, G., 2004, A&A 425, 973
80. Mouschovias, T., & Spitzer, L. 1976, ApJ 210, 326
81. Netzer, N., Elitzur, M. 1993, ApJ 410, 701
82. Nisini, B. et al. 2006, A&A 441, 159
83. Palla, F., Stahler, S.W. 1992, ApJ 392, 667
84. Panagia, N. 1991: Ionized winds from young stellar objects. In *The Physics of Star formation and early stellar evolution*, NATO Science Series Vol. 342, eds. C.J. Lada and N.D. Kylafis (Kluwer, Dordrecht), p. 565
85. Pesenti, N. et al. 2004, A&A 416, L9
86. Pudritz, R.E, Norman, C.A. 1983, ApJ 274, 677
87. Pudritz, R.E., Ouyed, R., Fendt, C. and Brandenburg, A. 2006: Disk Winds, Jets, and Outflows: Theoretical and Computational Foundations. In *Protostars & Planets V*, ed. by B. Reipurth, D. Jewitt, K. Keil (University of Arizona Press, Tucson), pp. 277–294
88. Raga, A.C., Cabrit, S., Dougados, C., Lavalley, C. 2001, A&A 367, 959
89. Ray, T.P. et al. 2006: Toward resolving the outflow engine: an observational perspective. In *Protostars & Planets V*, ed. by B. Reipurth, D. Jewitt, K. Keil (University of Arizona Press, Tucson), pp. 231–244
90. Reipurth, B., Bally, J. 2001, ARA&A 37, 403
91. Reipurth, B., Hartigan, P., Heathcote, S., Morse, J.A., Bally, J. 1997, AJ 114, 757
92. Reipurth, Heathcote, S., Yu, K.C., Bally, J., Rodriguez, L.F. 2000, ApJ 534, 317
93. Richer, J.S., Shepherd, D.S., Cabrit, S., Bachiller, R., Churchwell, E. 2000: Molecular Outflows from Young Stellar Objects. In *Protostars and Planets IV* eds Mannings, V., Boss, A.P., Russell, S.S. (University of Arizona Press, Tucson) pp. 867–894
94. Rodriguez, L.F. 1995, RevMexAA, Serie de conferencias, Vol. 1, p. 1
95. Sauty, C., Tsinganos, K. 1994, A&A 287, 893
96. Shang, H., Shu, F.H., Glassgold, A.E. 1998, ApJ 493, L91
97. Shang, H., Li, Z.-Y., Hirano, N. 2006: Jets and Bipolar Outflows from Young Stars: Theory and Observational Tests. In *Protostars & Planets V*, ed. by B. Reipurth, D. Jewitt, K. Keil (University of Arizona Press, Tucson), p. 261, 276
98. Shu, F.H., Najita, J., Ostriker, E.C., et al. 1994, ApJ 429, 781

99. Shu, F.H., Najita, J., Ostriker, E.C., Shang, H. 1995, ApJ 455, L155
100. Shu, F.H., Najita, J.R., Shang, H., Li, Z.-Y. 2000: Xwinds, Theory and Observations. In *Protostars & Planets IV*, ed. by V. Mannings, A.P. Boss, S.S. Russell, (University of Arizona Press, Tucson), pp. 789–813
101. Smith, M.D. 1986, MNRAS, 223, 57
102. Stapelfeldt, K. et al. 2003, ApJ 589, 410
103. Stone, J.M., Norman, M.L. 1992, ApJ, 389, 297
104. Tomisaka, K. 1998, ApJ 502, L163
105. Tenorio-Tagle, G., Cantó, J., Rozyczka, M. 1988, A&A 202, 256
106. Woitas, J., Ray, T.P., Bacciotti, F., Davis, C.J., Eisloeffel, J. 2002, ApJ 580, 336
107. Weber, E.J., Davis, L., Jr. 1967, ApJ 148, 217
108. Wu, Y. et al. 2004, A&A 426, 503

Star-disk Interaction in Classical T Tauri Stars

Silvia H. P. Alencar

Departamento de Física - ICEx-UFMG, CP. 702, Belo Horizonte, MG, 31270-901,
Brazil
silvia@fisica.ufmg.br

1 Introduction

Stars are born in molecular clouds. The mechanisms involved in the transformation of a cloud core into a star and the various evolutionary phases that the star-disk systems go through have been the center of attention of many recent studies (see recent reviews from the Protostars and Planets V conference [49]).

The protostar and disk are formed deeply embedded in their original collapsing cloud. From there on, in the case of low-mass stars, most of the stellar mass will be acquired through accretion from the disk. Initially, material still falls from the surrounding envelope to the disk, while the disk accretes

S. H. P. Alencar: *Star-disk Interaction in Classical T Tauri Stars*, Lect. Notes Phys. **723**,
55–73 (2007)
DOI 10.1007/978-3-540-68035-2_3

material into the star. At the same time, winds, jets and molecular outflows are driven by the young star-disk system and help dissipating the surrounding cloud. When the Pre-Main Sequence (PMS) star becomes optically visible, it is called a T Tauri star (TTS) if its spectral type is later than F or a Herbig Ae/Be star (HAeBe) if it has spectral type between B and F. At that point, the star still accretes from the disk, but at lower accretion rates (typically $\dot{M}_{acc} \sim 10^{-8}$ M$_\odot$ yr^{-1}) than in the protostellar phase.

The gas in the circumstellar disk will eventually vanish, but it is not actually clear what is the main cause of the gas dissipation. It can be just mostly accreted to the star, part of it may go into giant planets, it can also be disrupted by tidal instabilities and gaps created by planet formation or dissipated by UV photoionization by the central star. Some T Tauri stars, known as weak-lined T Tauri stars do not exhibit accretion disks at very early ages ($\sim$ 1 Myr) while others, the Classical T Tauri Stars (CTTSs), at the same ages and spectral types still do. Being brief or lasting longer, the star-disk interaction has a significant impact on early stellar evolution. It provides mass through accretion, helps regulate the angular momentum transfer and may be in part responsible for winds and jets observed in young stars.

In this contribution we will focus on accretion processes related to CTTSs. In Sect. 2, we will review the past and present models used to explain the observed characteristics of CTTSs. We will discuss the recent magnetic field measurements in Sect. 3 and the spectral diagnostics of magnetospheric accretion and accretion powered winds (Sect. 4). We will also emphasize in Sect. 5 the dynamical aspect of magnetospheric accretion and show how the models and simulations are dealing with such a characteristic.

2 Classical T Tauri Star Models: Past and Present

Classical T Tauri stars are young, optically visible, low mass stars that show signs of accretion from a circumstellar disk. They are a few million years old at most and still contracting down their Hayashi tracks. CTTSs show Li I in absorption, which is a telltale of youth for low mass stars. They present broad permitted emission lines and they are spectroscopic and photometrically variable. They show ultraviolet (UV), optical and infrared (IR) excesses with respect to the photospheric flux. They have strong magnetic fields ($\sim$2 kG) and are X-ray emitters. Over the years, as the observational characteristics above were discovered, several models were proposed to explain the CTTS phenomenon.

2.1 Boundary Layer Models

Lynden-Bell and Pringle (1974) [36] presented the idea that viscous dissipation in disks might be responsible for the IR excess. They also proposed that the UV excess could be produced in a shear equatorial boundary layer between the rapidly rotating disk (at more than 200 km s^{-1} just above the photosphere) and the slowly rotating star ($v \sin i \sim 10$ km s^{-1}, typically). In the boundary layer, the accreting material loses its kinetic energy in a series of shocks and emits a luminosity comparable to that released in the rest of the accretion disk, but over a very small area. Therefore the boundary layer must be much hotter than the rest of the disk, and so it will radiate at shorter wavelengths, peaking in the UV. It was only in the 80's that the idea became popular with the observational advances in UV and IR.

However, the boundary layer model (BLM) soon faced several observational challenges that it could not overcome. These models could not explain the inverse P Cygni profiles commonly observed in CTTS permitted emission lines, which show blueshifted emission peaks and redshifted absorption components at 200–300 km s^{-1} reaching below the continuum. The redshifted absorption component is thought to come from material that is receding from us towards the star at near free-fall velocities. In the BLM, material is supposed to slow down from the disk to the star, producing small radial velocities.

The BLM also could not explain the spectral energy distribution (the observed flux distribution as a function of wavelength) of many CTTSs that are only fitted by disk models without the inner disk regions (a disk hole). In the BLM, the disk is expected to continue down to the stellar photosphere.

These models had also difficulties to explain why CTTSs are generally found to be slow rotators ($v \sin i \sim 10$ km s^{-1}), since the accretion process should only add angular momentum and therefore spin-up the star.

2.2 Wind Models

Wind models were also suggested to explain the CTTSs emission line spectra [23, 24, 44]. These models tended to predict mostly P Cygni profiles (i.e. redshifted emission peaks and blueshifted absorption often going below the continuum) for the main emission lines. Consequently, they had a hard time generating the rather symmetric emission line profiles with blue-shifted features that do not reach the continuum, which are the most commonly observed in the main emission lines of CTTSs. The early wind models also tended to predict upper Balmer lines with strong wind absorption components and could not explain the spectra of many CTTSs that show these lines with very faint or even without blueshifted absorption. Wind models could not produce the observed inverse P Cygni profiles either.

All these arguments against the wind models do not mean that there is no wind in CTTSs, or no wind contribution to the observed profiles of CTTSs. It just indicates that the wind is not the main responsible for the permitted emission line profiles. The narrow blueshifted absorptions, however, remain wind features. They are attributed to cool outflowing gas that is further from the star than the accretion dominated emission.

2.3 Magnetospheric Accretion Models

Nowadays magnetospheric accretion models are the consensus to explain the many observed characteristics of CTTSs [25, 40, 41, 53]. These models are based on the idea of circumstellar disk accretion onto a magnetized young star. It is assumed that the stellar magnetic field is predominantly dipolar on the large scale and that it is strong enough to disrupt the disk at the so-called truncation radius or magnetopause, typically a few stellar radii from the star, where the magnetic pressure overcomes the ram pressure due to accretion. This forms a magnetospheric cavity (Fig. 1).

The accreting material that goes spiraling down the disk eventually reaches the inner disk region. If it is sufficiently ionized, its motion will then be controlled by the magnetic field. Part of it is ejected in a wind, and part is accreted onto the star along magnetic field lines, forming accretion columns.

As the near free-falling material hits the stellar surface, a strong accretion shock (a hot spot) is produced. The shock emission is responsible for the UV and optical continuum excess, known as veiling. The broad emission lines come from the accelerated material in the accretion columns and forbidden emission lines are formed in the wind. The IR excess comes from the combination of viscous dissipation and reprocessing of stellar and/or accretion radiation by the disk.

For accretion to occur, the truncation radius must be smaller than the corotation radius R_{co} (radial distance in the disk where the keplerian angular velocity matches the stellar angular velocity). Only inside R_{co} will the net force allow the material to accrete. Outside R_{co} the stellar angular velocity is greater than the Keplerian velocity, so that any material there which becomes

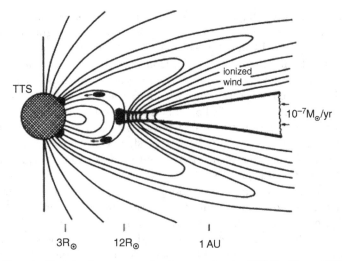

Fig. 1. Sketch of magnetospheric accretion in CTTSs (from [10])

locked to the stellar field lines will experience a centrifugal force that tries to throw the material away from the star (disk-wind).

This is the basic concept of magnetospheric accretion in CTTSs. So to begin with, we need strong stellar magnetic fields.

3 Magnetic Fields Measurements: Theory vs. Observations

Traditional magnetospheric accretion theories suggest that, when the star-disk system is in equilibrium, the stellar rotation rate will be determined by the Keplerian rotation rate near the truncation radius. The star is locked to the disk and these are therefore often called disk locking theories. Under the assumption that such an equilibrium situation exists, Königl (1991) [33], Collier Cameron & Campbell (1993) [11] and Shu et al. (1994) [53] have analytically analysed the interaction of a dipolar stellar magnetic field aligned with the stellar rotation axis and the inner accretion disk. Johns-Krull et al. (1999b) [29] obtained analytical expressions for the magnetic field strength in terms of the stellar mass and radius, the rotation period of the star and the mass accretion rate (see (2), (3) and (4) of [29]). Since these quantities have been estimated for several CTTSs, the models can then predict stellar magnetic field values that can later be compared with values obtained from the observations (Fig. 3).

Magnetic field measurements most of the time make use of the Zeeman splitting effect. When an atom is in the presence of a magnetic field, a spectral line is shifted in three components: two σ components, split to either side of the line center and a π component, which is unshifted. The wavelength shift of a σ component is given by:

$$\Delta\lambda = \frac{e}{4\pi m_e c^2}\lambda^2 gB$$

As shown above, the Zeeman shift has a λ^2 dependence, while Doppler line broadening mechanisms have just a λ dependence, so it is a good idea to look for magnetically sensitive lines in the IR in order to enhance the Zeeman effect compared to Doppler broadening. Making use of this characteristic, Valenti & Johns-Krull (2004) [58] fitted synthetic line models with magnetic field to magnetically sensitive Ti I lines in the IR spectra of CTTSs. An example of their analysis is shown in Fig. 2 for the CTTS BP Tau. The model with no magnetic field (single line) does not fit the data, while the model with a distribution of magnetic fields in the stellar surface, going from zero up to 6 kG, with a mean field value of 2.1 kG, gives a very good fit to the observations. In order to constrain non-magnetic broadening mechanisms Valenti & Johns-Krull (2004) [58] also fitted various CO lines that lie near the Ti I lines at 2.3 microns and have negligible Landé factors, which means they are Zeeman insensitive. As can be seen in the bottom panel of Fig. 2, the CO lines are

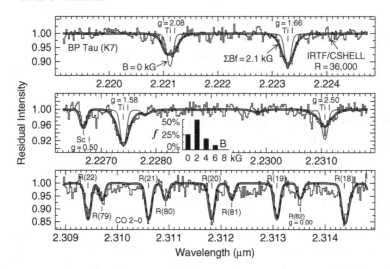

Fig. 2. An IR spectrum of the CTTS BP Tau (histogram) compared with synthetic line models with magnetic fields (*doubled curve*) and without (*single curve*). The inset histogram shows the distribution of magnetic fields used over the entire surface in the magnetic models to reproduce the observed spectrum. The Landé-g factor is given for each atomic line. Figure taken from [58]

much narrower than the Ti I lines and are well fitted by both the non-magnetic and the magnetic models, as expected. The surface magnetic field distribution of several TTSs has now been measured fitting the IR spectra with synthetic line models [29, 30, 32, 61].

The results presented in Fig. 2 show that the Zeeman broadening of magnetically sensitive lines depends on the distribution of magnetic field strengths. However they have limited sensitivity to magnetic geometry. On the other hand, circular polarization measurements for individual spectral lines are sensitive to magnetic geometry, but they provide limited information about the field strength.

The idea to use circular polarization is that, when viewed along the axis of a magnetic field, the Zeeman σ components are circularly polarized, but with opposite helicity, while the π component is absent. If one magnetic polarity is dominant at the visible surface of the star, net circular polarization is present in Zeeman sensitive lines. In that case, the magnitude of the wavelength shift is proportional to the surface average line of sight component of the magnetic field.

Most magnetospheric accretion models assume that the magnetic fields of TTSs are dipolar. It is true that the higher order multipoles fall off more rapidly than the dipole, so that at the inner edge of the disk, typically at a few stellar radii from the star, the dipole is likely to dominate. But close to the stellar surface, the magnetic field geometry is probably more complicated, with the presence of many multipole components. In support to that idea is the fact that most spectropolarimetric studies of cool stars do not clearly detect circular polarization in photospheric absorption lines (that form all over the surface of the star), within an upper limit of about 100 G [8, 59].

However, Johns-Krull et al. (1999a) [29] detected circular polarization in CTTSs emission lines that form predominantly in the accretion shock. There we are not looking at the whole stellar surface, but we are isolating a small portion of it. The strongest circular polarization signal appears in the narrow emission component of the He I line (5876 Å). In contrast to the photospheric features, which show no net field polarity in the line of sight, the narrow helium emission arises in a region with mean line of sight fields from 1–2.75 kG that exhibit a high degree of organization, indicating that a single polarity dominates in the He I formation region. Valenti & Johns-Krull (2004) [58] have further measured the line of sight component of the magnetic field on six consecutive nights for four stars. The measured values vary smoothly on rotational timescales, implying a lack of symmetry about the rotation axis in the accretion or the magnetic field, or both. Simple models consisting of a single spot at a given latitude that rotates with the star fit the data quite well.

The complicated surface topology of magnetic fields in CTTSs results in no net polarization in photospheric absorption lines. However, the accreting material apparently follows the dominant polarity of the field at the inner disk edge, so that emission lines formed in the accretion shock preferentially illuminate a dominant polarity component of the field, producing substantial circular polarization in these emission lines. The dominant polarity component near the truncation radius is expected to be the dipole component, as predicted by the magnetospheric theories, if the truncation radius is far enough for it to dominate over the others. However, Gregory et al. (2006) [21] and Jardine et al. (2006) [26] have shown that accretion is likely to occur along non-dipolar field lines.

In Fig. 3 are shown the most recent compilation of mean magnetic field intensities measured by Johns-Krull and collaborators together with the predicted field strengths from Shu et al. (1994) [53]. We can readily see that the mean fields do not agree with the theory. A hint to why we do not see such a correlation may come from the field topology measurements, since they indicate the magnetic field on the TTSs surface is not globally dipolar. The dipole component may then actually be much less important than predicted by the theories. There is therefore the need to include non-dipole fields in the magnetospheric accretion theory to try to reconcile theory and observations. This was done by Johns-Krull & Gafford (2002) [31] who compared the relationships predicted by magnetospheric accretion theories among stellar mass, radius, rotation period and disk accretion rate and the observed relations from several studies of CTTSs. They were only able to reproduce the observed correlations after including non-dipole field topologies in the model of Ostriker & Shu (1995) [45].

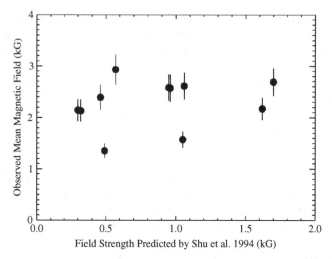

Fig. 3. Observed mean magnetic field strength as a function of the predicted field strength for the theory of Shu et al. (1994) [53]. There is no statistically significant correlation between the observed and predicted magnetic field strengths. Figure taken from [7]

4 Spectral Diagnostics of Magnetospheric Accretion

4.1 Emission-line Profiles

The magnetic field distribution and topology is one of the most important ingredients to understand and correctly describe the magnetospheric accretion process in CTTSs. There are, however, several other diagnostics of magnetospheric accretion, such as the permitted emission line profiles of CTTSs that are supposed to be formed in the accretion funnel.

These emission line profiles show a wide variety of morphologies (symmetric, double-peaked, P Cygni, inverse P Cygni) and they can vary from one type to another in a single star. In common to all types of profiles is the broad line width with several 100 km s^{-1} indicative of bulk motion of the circumstellar material. The emission line profiles are important because they encode both geometrical and physical information on the accretion process and its rate. So a straightforward idea is to use them to test and refine the magnetospheric accretion models.

The standard magnetospheric accretion models assume an axisymmetric geometry in which the density structure along the funnel is calculated using a steady mass accretion rate and free-fall along dipolar field lines, which come from a geometrically thin but optically thick disk at a range of radii inside the co-rotation radius. It is generally assumed that the kinetic energy of the accreting material is completely thermalized.

The radiative transfer calculations are typically performed with the Sobolev approximation i.e. assuming that the contributions to a spectral line formation at any given frequency occur locally, that is valid under the assumption of high velocity gradients. The temperature structure, which is a significant parameter, is unfortunately poorly constrained. A simple adjustable volumetric heating rate combined with a schematic radiative cooling rate, which leads to a temperature structure that goes as the reciprocal of the density is often adopted [25, 40, 41, 56]. The magnetospheric accretion models may also include rotation and line damping.

Actually the models are able to compute a huge variety of profiles, as observed, and the calculations are performed for several different atomic species, such as hydrogen, sodium, calcium and oxygen. Several hydrogen transitions in the optical and IR have also been calculated.

Even with all the simplifications and assumptions that are made in the models, it is fair to say that the magnetospheric accretion models do a very good job in reproducing the main observed characteristics of the permitted

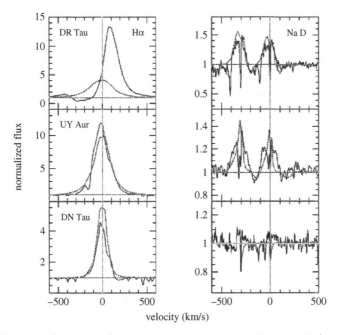

Fig. 4. Observed (*solid lines*) and magnetospheric accretion model (*dotted lines*) permitted emission line profiles of three CTTSs. We see that the magnetospheric accretion models reproduce very well the emission part of the profiles but do not reproduce the blueshifted absorptions in Hα and NaD of any of the stars nor the shape and intensity of the Hα line of a high mass accretion rate CTTS like DR Tau. Those features are believed to be due to winds. Figure taken from [41]

emission line profiles (see Fig. 4), such as the broad line widths and the occasional inverse P Cygni profiles [40, 41].

The models are also often able to reproduce several lines simultaneously, which allows to better constrain the many fitting parameters. Care must be taken to determine the accretion and magnetospheric parameters with only one line like Hα, which is the most commonly used, since it is vulnerable to have contributions from outflows (see [2, 3]) and it may also be significantly spatially extended.

Lately the magnetospheric accretion model has been used to determine mass accretion rates across the mass spectrum, from Herbig Ae stars, UX Ori being a good example [42] and also to very low mass objects and brown dwarfs [43].

In general, the fit to the emission component of the observed line profiles is quite good but there are exceptions like DR Tau, which is a high mass accretion rate system (see Fig. 4). In that case, it has been suggested by Muzerolle et al. (2001) [41] that the wind can give an important contribution to the emission line profile.

4.2 Accretion Driven Winds

It is widely accepted that winds are important ingredients of star formation, but the wind contribution to the line profiles of CTTSs has not been very much studied in the last years, because of the apparent importance of accretion to the observed profiles.

Alencar et al. (2005) [2] have recently shown that the observed emission profiles of RW Aur, a high mass accretion rate CTTS, can have a strong wind contribution, as predicted by Muzerolle et al. (2001) [41] for such star-disk systems. In Fig. 5 we show the magnetospheric accretion and disk wind profiles compared with mean Hα and Hβ observations of RW Aur. We can see that the disk wind models fit better the observations, indicating that, in this case, the wind contribution is important. Alencar et al. (2005) [2] calculated the magnetosphere and wind models separately, while to be consistent, the accretion and wind contributions must be calculated at the same time, as Kurosawa et al. (2005, 2006) [34, 35] have recently done for the first time. They computed hybrid models where both effects are taken into consideration and they are able to reproduce all types of emission line profiles observed in CTTSs (see Fig. 5 of [7]). Only hybrid models are able to simultaneously reproduce the emission and the narrow blueshifted absorption, which is attributed to cool outflowing gas that is further from the star than the accretion dominated emission. The narrow blueshifted absorption features exhibit a great variety in depth, width and velocity from one star to another, although there is a general tendency for them to be more prominent in stars with high mass accretion rates [14], indicating that these are not purely stellar winds but are powered by an accretion/outflow connection.

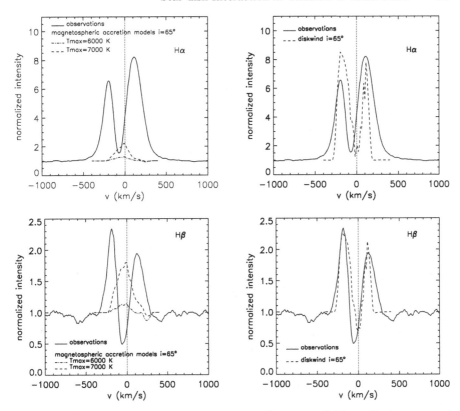

Fig. 5. RW Aur mean observed line profiles are shown as solid lines. **Left**: magnetospheric accretion profiles (*dashed and dash-dotted lines*). **Right**: disk-wind profiles (*dashed lines*). From Alencar et al. (2005)

Recently another spectral wind diagnostic has been identified with the confirmation of the presence of a hot helium wind among stars with high accretion and outflow rates provided by profiles in the near infrared triplet line of He I 10830 Å [15].

In Fig. 6, kindly provided by Suzan Edwards prior to publication, we see the He I 10830 Å line of three CTTSs (DR Tau, AS 353 A, DO Tau) and one WTTS (V819 Tau). The CTTSs present very impressive P Cygni profiles with blueshifted absorptions going way below the continuum in a spectral region where the continuum is mostly due to the star. At 1 micron the accretion shock does not contribute to the continuum and the IR excess from the disk is still rising and does not contribute much either. In Fig. 6 we can see that for DR Tau, AS 353A and DO Tau the blueshifted absorption reaches then into the stellar continuum over a continous and broad velocity range going from near the stellar rest velocity to the terminal wind velocity, as traced by the forbidden emission lines. In order to have helium emission we further need high temperatures (T > 10 000 K). So the acceleration region of the helium

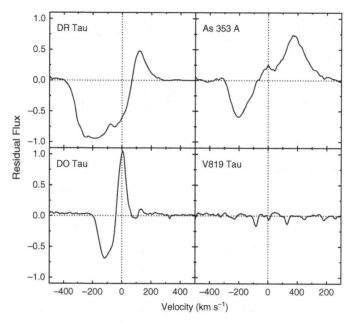

Fig. 6. Observed He I 10830 Å line profiles of 3 CTTSs (DR Tau, AS 353 A, DO Tau) and 1 WTTS (Suzan Edwards, private communication)

wind is close to the star, occulting a significant portion of it, in a region of high temperature.

One possibility is that the helium wind comes from polar/coronal regions. This would allow magnetospheric accretion funnels to co-exist with such a hot wind. It would actually be difficult for disk wind material to absorb the radiation from the stellar disk at all velocities, from the rest to the terminal wind speed. As extreme examples, TW Hya and T Tau are pole on CTTSs and they also present the He I (10830 Å) line with the same strong and extended blueshifted absorptions (see [12]). In those cases it is almost impossible for an inner wind to be of disk origin and produce such absorptions of the stellar continuum.

If the wind arises from the stellar coronae it would not be a normal stellar wind but rather an accretion powered stellar wind, since this wind is strongest in stars with the highest mass accretion rates. Low mass accretion rate CTTSs, like AA Tau, show He I (10830 Å) profiles that are compatible with a disk wind with narrow blueshifted absorptions much like in Hα (Suzan Edwards, private communication) and WTTSs do not show any absorption or emission at all, as can be seen in Fig. 6 for V819 Tau.

The picture of the star-disk interaction in the PMS that is emerging from observational results includes magnetospheric accretion, a disk wind, and an accretion powered stellar wind, as discussed by Matt & Pudritz (2005) [39] and by Ferreira et al. (2006) [18] and shown schematically in Fig. 1.

4.3 Accretion Shock Evidences

Another key parameter of the magnetospheric accretion model is the presence of accretion shocks that create hot spots at the stellar surface where the infalling material reaches the star. Hot spots were initially inferred from rotational modulation of light curves by Bouvier et al. (1995) [4]. In order to ensure that the photometric modulation was mainly caused by hot spots and not by cold ones, the temperature and minimum size of the spots responsible for the stellar brightness modulation were derived by comparing the observed amplitude of variability in different wavelengths with hot and cool spot models calculated with various temperatures and spot sizes. Hot spot models failed to reproduce the observations of WTTSs, while both hot and cool spots can be responsible for the observed modulation of CTTSs. Half of the observed CTTSs in Bouvier et al. (1995) [4] had amplitudes consistent with a modulation by cool spots, and half by hot spots. Their conclusion was that CTTSs present both cool and hot spots, while WTTSs only show evidence for cool spots.

Another evidence for accretion shocks comes from the successful fitting of the UV and optical excesses observed in CTTSs by accretion shock models ([9, 22]). The shape of the excess can be understood as optically thick emission from the heated photosphere below the shock and optically thin emission from the preshock and postshock regions. According to the models, the accretion shock will initially have $T \sim 10^6$ K and emit in soft X-rays, which are reprocessed both by the accretion stream and the stellar photosphere, accounting for the optical and UV continuum emission. An important product of such a fitting is the reliable determination of mass accretion rates, which can later be used to calibrate other accretion diagnostics that are easier to measure than the UV excess, such as the luminosity of emission lines like Paschen β.

4.4 X-ray Emission

Young stars are bright in X-rays from Class I to Class III stages. They can be up to 10^4 times more luminous than the Sun in X-rays, which are thought to be mainly produced by magnetic reconnection flares [16].

In the magnetic reconnection, oppositely directed magnetic field lines are brought together into a strong eddy that pushes away the gas. The magnetic reconnection changes the topology of magnetic fields by breaking magnetic field lines and reconnecting them in a different way. In doing so, it can liberate magnetic energy into other forms such as kinetic energy, heat and light.

Recently a collaboration known as the Chandra Orion Ultradeep Project (COUP) led by Eric Feigelson (Penn State Univ.) has obtained 13 days of observations of the Orion Nebular Cluster with the Chandra X-ray observatory (ApJS Volume on the COUP project [17, 47]). They confirmed that there is no X-ray quiet young population and that the X-rays from young stars are mainly of enhanced solar-like coronae origin, but they can also have contributions of soft X-rays from accretion columns and of intense flares from the reconnection of field lines taking part in the magnetic star-disk interaction. Such intense flares are predicted by numerical simulations of the star-disk interaction, where differential rotation between the stellar field lines and the inner disk are shown to lead to the lines expansion, opening and reconnection (see Sect. 5.1).

CTTSs are known to be less X-ray active than WTTSs but this difference is not due to the fact that CTTSs rotate more slowly than WTTSs, since in the PMS no activity-rotation correlation is observed, while it is in the later phases of stellar evolution [47]. The reason for such a difference is still under debate and could be due to X-ray absorption by circumstellar features around CTTSs. Accretion could also change the magnetosphere itself, loading the field lines and therefore not allowing the high density plasma to reach very high temperatures needed to emit X-rays. Another possibility under discussion is that accretion may change the stellar structure, inhibiting the dynamo process and therefore affecting the rising of magnetic field lines at the stellar surface [47].

5 The Future of CTTS Models

In terms of observational evidences, we can say that magnetospheric accretion is overall in a robust ground. We have seen that strong magnetic fields are present in CTTSs [58]. It was shown that accretion columns are inferred through the observation of inverse P Cygni profiles with redshifted absorptions at several hundred km s^{-1} [13], which indicates that the gas is accreted onto the star from a distance of a few stellar radii and we saw that emission lines are formed, at least partially, in accretion funnel flows [25, 40, 41]. Accretion shocks are inferred from the rotational modulation of light curves by bright surface spots [4] and accretion shock emission models have also successfully reproduced the observed spectral energy distribution of optical and UV excess [9, 22].

However some caveats do apply to the models and there are aspects such as the variability of the accretion process that are not taken into account by the very good but steady state, axisymmetric accretion models.

The first and probably one of the most important problems is that the temperature structure of the accretion flow is essentially arbitrary, and this has a significant impact on the line source functions and consequently on the profiles themselves.

Martin (1996) [37] showed that in the funnel, the heating is dominated by adiabatic compression and the cooling comes mainly from free-free emission (Bremsstrahlung radiation) and line emission from ions such as CaII and MgII. The temperature structure he obtained is quite different from the one commonly used in the magnetospheric accretion models. However, the somewhat arbitrary temperature law proposed by Hartmann et al. (1994) [25] yields profiles that definitely look like the observations, while the Martin (1996) one, which was calculated consistently, does not. There is therefore the need to further analyse the heating and cooling processes that take place in the accretion funnel.

Another important point that the standard models do not deal with, is that the accretion process is very variable. Steady state axisymmetric models cannot obviously explain the observed variability.

SU Aur is a typical example of CTTS that shows variable emission lines [27, 46]. It also shows periodic infall and outflow signatures that are 180 degrees out of phase, indicating that the wind is strongest when the infall is weakest. This can be explained by a magnetospheric accretion model with an inclined dipole field with respect to the rotation axis (non-axisymmetric) in which accretion and outflow are naturally favored in opposite phases.

TW Hya is a pole-on CTTS that, despite of that, presents a daily variability of its emission line profiles [1]. In that case, even an inclined dipole cannot easily explain such variabilities and there is evidence therefore for non-steady accretion towards the star, a feature that the standard magnetospheric accretion models do not take into account.

Overall, the accretion/ejection processes appear to be dynamical on several timescales. It is variable over hours for non-steady accretion, like seen in the veiling variability of RU Lup during a single night [55]. In weeks, for rotational modulation, like SU Aur discussed before, and like the veiling variability of AA Tau [5]. These week-long variabilities are most likely due to an inclined dipole, causing accretion to be periodically favored at certain phases. Accretion is variable over months, for global instabilities of the magnetosphere. Due to differential rotation between the star and the disk, numerical models have shown that the magnetosphere is expected to expand, open and eventually reconnect [19, 20, 38]. It was also shown that this expansion can be measured observationally by the projected radial velocity of the redshifted absorption component of Hα, which is formed in the funnel flow [5]. This was observed for AA Tau, as the radial velocity of the redshifted component decreases in a few rotational periods. At the lowest radial velocity level, accretion is inhibited (inflated magnetosphere). The observed veiling then goes to zero and the flux in the lines too. A few days later (maybe reconnection) everything is back again. The accretion/ejection process is also observed to be variable in years from outbursts like EXORs and FU Ori, a recent example being the McNeil nebula [48].

5.1 Dynamical Models

The simplest extension to the standard magnetospheric accretion models, in order to investigate the rotational profile variability, is to brake the axisymmetry of the dipole, leaving, for example, curtains of accretion in azimuth, but still keeping the magnetic and rotation axis aligned. Such models were proposed to explain the variability of the observations [54] and are predicted by 3D MHD simulations [51]. Symington et al. (2005) [56] calculated such models and saw that the line profiles overall agree with the observations but they ended up with a variability much larger than usually observed in CTTSs. They suggest that the magnetosphere has probably a high degree of axisymmetry broken by higher density streams that produce the observed variability.

A more complicated but very interesting view of the star-disk interaction has recently emerged from MHD simulations. Romanova et al. (2002) [50] performed impressive MHD simulations of stellar magnetic field and disk interactions in quiescent regime. They calculated 2D MHD simulations of disk accretion to a rotating aligned dipole and showed that funnel flows, where matter flows out of the disk plane and essentially free-falls along the stellar magnetic field lines, are a robust feature of disk accretion to a dipole. In the simulations, which run up to more than 60 Keplerian periods near the inner disk truncation radius, the disk, represented by a "density structure", truncates and a funnel flow forms near the truncation radius, where the magnetic pressure of the dipole is comparable to the kinetic plus thermal pressure of the accreting material. Some outflow by centrifugal force is present in the simulations and seen to be quasi periodic.

Romanova et al. (2003) [51] also did 3D ideal MHD simulations of the disk accretion to a slowly rotating star with an inclined dipole magnetic field. In these simulations, matter is shown to accrete to the inclined dipole forming non-axisymmetric structures in the stellar magnetosphere. The simulations show that the flow of matter has different shapes at different density levels. The low density part of the flow covers almost the entire magnetosphere, while the larger density regions of the flow accrete in streams. The streams may obscure the light emitted from the stellar surface or from hotspots at the surface of the star, thereby causing stellar variability. The inner regions of the disk often become warped or tilted as the system evolves, as predicted by Terquem & Papaloizou (2000) [57].

Another result of the numerical simulations is the presence of a dynamical interaction between the disk and the stellar magnetosphere. Several simulations have shown that differential rotation along the field lines between the star and the inner disk region leads to the lines expansion, opening and reconnection [19, 20, 38, 52, 60]. When the magnetic field reconnects, strong X-ray flares can be produced. After the reconnection the initial magnetospheric configuration is restored. The timescale for this to occur is of a few rotation periods, determined by the diffusion of the magnetic flux through the inner regions of the disk. That is apparently what was detected in the AA Tau observing campaigns by Bouvier et al. (2003, 2005) [5, 6], but those are the only datasets that allowed to investigate these results. There is therefore the need for longer observing campaigns of CTTSs combining photometry, spectroscopy and polarimetry to investigate the time variability of the magnetospheric accretion and the star-disk interaction in a timescale of weeks to months.

6 Conclusions

The general picture that is emerging nowadays of a CTTS is that of an interacting star-disk system composed of a young magnetized star and its circumstellar disk. The young star has a strong magnetic field that presents a complex topology at the surface and interacts with the circumstellar disk through dipole or multipole field lines, depending whether the disk truncation radius is close to the star (multipoles) or far enough from the star for the dipole to dominate. Magnetospheric accretion, with the overall characteristics predicted by magnetospheric accretion models, is a robust feature in CTTSs, supported nowadays by several observational evidences such as the presence of strong magnetic fields (~ 2 kG), the presence of accretion shocks, the observation of redshifted absorption from infalling material near free-fall velocities and strong emission lines formed in accretion columns. Accretion also seems to be the driving source of hot stellar winds and disk winds, since such wind signatures are prominent in stars with high mass accretion rates. Disk Winds, accretion-powered stellar winds and magnetospheric accretion apparently coexist in CTTSs. Accretion and ejection are then strongly related processes and models that try to explain CTTSs should take both into account.

The star-disk interaction is very dynamical on several timescales (from hours to weeks, months and years) and is mediated by the stellar magnetic field, as evidenced by synoptic studies of CTTSs and by the recent MHD simulations of the star-disk interaction. There is nowadays the need for month-long observational campaigns of CTTSs in order to better understand the dynamics of the star-disk interaction and to test the predictions of the MHD simulations.

Acknowledgements

I thank the organisers for a very enjoyable school and JETSET and CNPq (grant 201228/04-1) for financial support. I would also like to thank Suzan Edwards for Fig. 6 and many explanations on accretion driven stellar winds, and the PPV team (Jérôme Bouvier, Chris Johns-Krull, Tim Harries and Marina Romanova) for the discussions that, in most part, lead to this contribution.

References

1. S.H.P. Alencar, C. Batalha: ApJ, 571, 378
2. S.H.P Alencar, G. Basri, L. Hartmann, & N. Calvet: A&A, 440, 595
3. I. Appenzeller, C. Bertout, & O. Stahl: A&A, 434, 1005 (2005)
4. J. Bouvier, E. Covino, O. Kovo, et al.: A&A, 299, 89 (1995)
5. J. Bouvier, et al.: A&A, 409, 169 (2003)

6. J. Bouvier, T. Boutelier, S.H.P. Alencar, C. Dougados, C.: Magnetospheric Accretion-Ejection Processes in the CTTS AA Tau. In: *Protostars and Planets V*. LPI Contribution No. 1286., p. 8150
7. J. Bouvier, S.H.P. Alencar, T.J. Harries, C.M. Johns-Krull, M.M. Romanova: Magnetospheric accretion in classical T Tauri stars. In: *Protostars and Planets V*, ed by B. Reipurth, D. Jewitt, K. Keil (University of Arizona Press, Tucson), pp. 479–494 (2006)
8. E.F. Borra, G. Edwards, M. Mayor: ApJ, 284, 211 (1984)
9. N. Calvet, E. Gullbring: ApJ, 509, 802 (1998)
10. M. Camenzind: Reviews of Modern Astronomy, 3, 234 (1990)
11. A. Collier Cameron, & C.G. Campbell: A&A, 274, 309 (1993)
12. A.K. Dupree, N.S. Brickhouse, G.H. Smith, J. Strader: ApJ, 625, L131 (2005)
13. S. Edwards, P. Hartigan, L. Ghandour, C. Andrulis: AJ, 108, 1056 (1994)
14. S. Edwards: Ap&SS, 287, 47 (2003)
15. S. Edwards, W. Fischer, J. Kwan, L. Hillenbrand, L., A.K. Dupree: ApJ, 599, L41 (2003)
16. E.D. Feigelson, T. Montmerle: ARA&A, 37, 363 (1999)
17. E.D. Feigelson, et al.: ApJS, 160, 379 (2005)
18. J. Ferreira, C. Dougados, S. Cabrit: A&A, 453, 785 (2006)
19. A.P. Goodson, R.M. Winglee, K.-H. Boehm: ApJ, 489, 199 (1997)
20. A.P. Goodson, R.M. Winglee: ApJ, 524, 159 (1999)
21. S.G. Gregory, M. Jardine, I. Simpson, J.-F. Donati: MNRAS, 371, 999 (2006)
22. E. Gullbring, N. Calvet, J. Muzerolle, L. Hartmann: ApJ, 544, 927 (2000)
23. L. Hartmann, S. Edwards, E. Avrett: ApJ, 261, 279 (1982)
24. L. Hartmann, N. Calvet, E. Avrett, R. Loesler: ApJ, 349, 168 (1990)
25. L. Hartmann, R. Hewett, N. Calvet: ApJ, 426, 669 (1994)
26. M. Jardine, A.C. Cameron, J.-F. Donati, S.G. Gregory, K. Wood: MNRAS, 367, 917 (2006)
27. C.M. Johns, G. Basri: ApJ, 449, 341 (1995)
28. C.M. Johns-Krull, J.A. Valenti, C. Koresko: ApJ, 510, L41 (1999a)
29. C.M. Johns-Krull, J.A. Valenti, C. Koresko: ApJ, 516, 900 (1999b)
30. C.M. Johns-Krull, J.A. Valenti, S.H. Saar, A.P. Hatzes: ASP Conf. Ser.: Magnetic Fields Across the Hertzprung-Russell Diagram, ed by G. Mathys, S.K. Solanki, and D.T. Wickramasinghe. (Astronomical Society of the Pacific, San Francisco) 248, 527
31. C.M. Johns-Krull, A.D. Gafford: ApJ, 573, 685 (2002)
32. C.M. Johns-Krull, J.A. Valenti, S.H. Saar: ApJ, 617, 1204 (2004)
33. A. Koenigl: ApJ, 370, L39 (1991)
34. R. Kurosawa, T.J. Harries, N.H. Symington: Formation of H-alpha from Classical T Tauri Stars: The Disc, Wind, and Accretion Hybrid Model. In: *Protostars and Planets V*. LPI Contribution No. 1286., p. 8412, 8412
35. R. Kurosawa, T.J. Harries, N.H. Symington: MNRAS, 370, 580 (2006)
36. D. Lynden-Bell, & J.E. Pringle: MNRAS, 168, 603 (1974)
37. S.C. Martin: ApJ, 470, 537
38. S. Matt, A.P. Goodson, R.M. Winglee, K.-H. Bohm: ApJ, 574, 232
39. S. Matt, R.E. Pudritz: ApJ, 632, L135 (2005)
40. J. Muzerolle, L. Hartmann, N. Calvet: AJ, 116, 455 (1998)
41. J. Muzerolle, N. Calvet, & L. Hartmann: ApJ, 550, 944 (2001)
42. J. Muzerolle, P.D' Alessio, N. Calvet, & L. Hartmann: ApJ, 617, 406 (2004)

43. J. Muzerolle, K.L. Luhman, C. Briceño, L. Hartmann, N. Calvet: ApJ, 625, 906 (2005)
44. A. Natta, & C. Giovanardi: ApJ, 356, 646 (1990)
45. E.C., Ostriker, & F.H. Shu, 1995, ApJ, 447, 813
46. P.P. Petrov, E. Gullbring, I. Ilyin, et al.: A&A, 314, 821 (1996)
47. T. Preibisch, et al.: ApJS, 160, 401 (2005)
48. B. Reipurth, C. Aspin: ApJ, 606, L119 (2004)
49. B. Reipurth, D. Jewitt, K. Keil *Protostars and Planets V*, ed by B. Reipurth, D. Jewitt, K. Keil (University of Arizona Press, Tucson, 2007) XXIV, 951
50. M.M. Romanova, G.V. Ustyugova, A.V. Koldoba, R.V.E. Lovelace: ApJ, 578, 420 (2002)
51. M.M. Romanova, G.V. Ustyugova, A.V. Koldoba, J.V. Wick, R.V.E. Lovelace: ApJ, 595, 1009 (2003)
52. M.M. Romanova, G.V. Ustyugova, A.V. Koldoba, R.V.E. Lovelace: ApJ, 616, L151 (2004)
53. F. Shu, J. Najita, E. Ostriker, et al.: ApJ, 429, 781 (1994)
54. K. Smith, G.F. Lewis, I.A. Bonnell, J.P. Emerson: A&A, 378, 1003 (2001)
55. H.C. Stempels, N. Piskunov: A&A, 408, 693 (2003)
56. N.H. Symington, T.J. Harries, R. Kurosawa: MNRAS, 356, 1489 (2005)
57. C. Terquem, & J.C.B. Papaloizou: A&A, 360, 1031 (2000)
58. J.A. Valenti, & C.M. Johns-Krull: Ap&SS, 292, 619 (2004)
59. S.S. Vogt: ApJ, 240, 567 (1980)
60. B. von Rekowski, B.,A. Brandenburg: A&A, 420, 17 (2004)
61. H. Yang, C.M. Johns-Krull, J.A. Valenti: ApJ, 635, 466 (2005)

Part II

Magneto-Hydrodynamic Models

Introduction to Magneto-Hydrodynamics

Guy Pelletier

Laboratoire d'Astrophysique de Grenoble, BP 53, 38041 Grenoble Cedex, France
Université Joseph Fourier, Grenoble, France
Guy.Pelletier@obs.ujf-grenoble.fr

Abstract. Magneto-Hydrodynamics (hereafter MHD)describes plasmas on large scales and more generally electrically conducting fluids. This description does not discriminate between the various fluids that constitute the medium. In laboratory, it allows to globally describe a plasma machine, for instance a toroidal nuclear fusion reactor like a Tokamak. In astrophysics it plays an essential role in the description of cosmic objects and their environments, as well as the media, such as the interstellar or the intergalactic medium. A set of phenomena are specific to MHD description. Some of them will be presented in this lecture such as the tension effect, confinement, magnetic diffusivity, magnetic field freezing, Alfvén waves, magneto-sonic waves, reconnection. A celebrated phenomenon of MHD will not be introduced in this brief lecture, namely the dynamo effect.

G. Pelletier: *Introduction to Magneto-Hydrodynamics*, Lect. Notes Phys. **723**, 77–101 (2007)
DOI 10.1007/978-3-540-68035-2_4 © Springer-Verlag Berlin Heidelberg 2007

1 What is MHD?

Magneto-HydroDynamics (MHD) describes the dynamics of an electrically conducting continuous medium where the Lorentz force significantly acts under the following conditions:

- the electric neutrality is realized locally
- the dynamics is that of a single fluid
- Ohm's law applies.

This is never truly realized in a plasma, but these conditions are easily realized with a good approximation in a collisional plasma at scales much larger than the mean free path. These conditions are not obviously realized in hot plasmas and it is useful to look for criteria based on the main reference parameters: such that the typical intensity of the field B_0, the scale of its variations ℓ_0 and the plasma mass density ρ_0. One then defines the typical current density (the international system of units will be used):

$$J_0 = \frac{B_0}{\mu_0 \ell_0} \tag{1}$$

and the typical Lorentz force density:

$$\frac{B_0^2}{\mu_0 \ell_0} \ . \tag{2}$$

The Alfvén time τ_A and the Alfvén velocity V_A characterize the dynamics. Indeed a fluid element is accelerated by the Lorentz force such that

$$\rho_0 \frac{\ell_0}{\tau_A^2} \sim \frac{B_0^2}{\mu_0 \ell_0} \ ,$$

thus one defines

$$V_A \equiv \frac{\ell_0}{\tau_A} \equiv \frac{B_0}{\sqrt{\mu_0 \rho_0}} \ . \tag{3}$$

Non-relativistic MHD holds as long as the Alfvén velocity is much smaller than the velocity of light and thus the "displacement current" can be neglected in Ampère-Maxwell equation (magnetostatic approximation).

It is well known in plasma physics that local neutrality is approximately satisfied when the motions are slow compared to the electron thermal speed and developed on large scales compared to Debye length. Indeed under those conditions the electron fluid is in a quasi Boltzmann equilibrium and screens the ion charge, so that the deviation to local neutrality is on the order of λ_D^2/ℓ_0^2, where λ_D is the Debye length. However we will discuss this point in more details later on.

The MHD description (see text books by Parker [7], Priest [9], Goedbloed & Poedts [3]) is designed in order to be adequate for the description of the global motions of a medium at speeds much slower than the speeds of the particles belonging to the populations lighter than the most inert component; in other words the motions are comparable to the motions of the heaviest population (ions or, in a weakly ionized medium, neutral if they are coupled by collisions), the lighter populations (electrons and lighter ions), that have higher microscopic motions, globally follows in quasi equilibrium, their inertial force being statistically negligible. In collisional medium, it is sufficient that the Alfvén time be longer than the longest collisional time for fulfilling the single motions criteria. This amounts to assume that the scale ℓ_0 is larger than the larger mean free path. At these scales, all the species have a strong collisional coupling and move as a single fluid. In a weakly collisional plasma the slower microscopic time is the cyclotron period of the heaviest particles. The MHD description then requires that these individual motions be smoothed out which requires that $\omega_{ci}\tau_A \ll 1$ or equivalently $\ell_0 \gg r_0 \equiv \frac{V_A}{\omega_{ci}}$. This length r_0 is a kind of ion Larmor radius and the condition for MHD validity can be seen as the condition for the ions to be "magnetized" at the considered scale. This can also be expressed as the inertial length of the ions since $r_0 = \delta_i = c/\omega_{pi}$ (ω_{pi} being the ion plasma pulsation); at a scale smaller than δ_i the ion fluid tends to frozen. MHD is therefore the approximation of leading order with respect to the small parameter r_0/ℓ_0.

The consistence of MHD approximation is questionable since the existence of an electric current implies that electrons and ions are moving at different speeds. The required velocity difference can be estimated, assuming electroneutrality:

$$\frac{|\boldsymbol{u}_e - \boldsymbol{u}_i|}{V_A} \sim \frac{J_0}{n_e q_e} V_A \sim \frac{B_0 \sqrt{\mu_0 n_i m_i}}{\mu_0 \ell_0 n_e q_e B_0} \sim \frac{c}{\omega_{pi}\ell_0} \sim \frac{r_0}{\ell_0} . \tag{4}$$

As will be seen several times, this size r_0 defines the scale beyond which MHD description is valid.

2 Dynamical Equations

2.1 Local Conservation Laws of Matter and Electric Charge

As long as no chemical nor nuclear reaction is considered, each fluids preserves its mass in the flow, which is locally expressed by the continuity equation:

$$\frac{\partial}{\partial t} n_a + \operatorname{div} n_a \boldsymbol{u}_a = 0 . \tag{5}$$

The single fluid description requires that the total mass density $\rho \equiv \sum_a n_a m_a$ and the global fluid velocity $\boldsymbol{u}$ be linked by the law of matter conservation, which is exactly insured by defining this global fluid velocity as the motion of the center of mass of the fluid element

$$u \equiv \frac{\sum_a n_a m_a u_a}{\sum_a n_a m_a} \; ; \tag{6}$$

thus the mass transport equation reads:

$$\frac{\partial}{\partial t}\rho + \operatorname{div}\rho u = 0 \; . \tag{7}$$

In the particular case of a fully ionized plasma constituted by only one ion population (H^+ for instance), the definition of the barycentric velocity and the definition of the current density, $J \equiv \sum_a n_a q_a u_a$, constitutes a system of two linear equations that can be easily inverted and the result can be expanded in powers of the small mass ratio m_e/m_i. The following result is obviously obtained:

$$u_i = u(1 + \mathcal{O}(m_e/m_i)) \text{ and} \tag{8}$$
$$u_e = (u + J/n_e q_e)(1 + \mathcal{O}(m_e/m_i)) \; . \tag{9}$$

The condition for a weak velocity difference previously given insures that the difference of u_e with u is relatively weak.

Momentum Transport

The equation of motions for each fluids reads:

$$\frac{\partial}{\partial t} n_a m_a u_{a,i} + \nabla_j (\rho u_{a,i} u_{a,j} + P_{a,ij}) =$$
$$n_a q_a (E_i + \varepsilon_{ijk} u_{a,j} B_k) + n_a m_a g_i + \sum_b \Pi_{ab,i} \; , \tag{10}$$

where Π_{ab} is the momentum lost by the particle of type "a" through elastic collisions with particles of type "b"; and one has $\sum_{ab} \Pi_{ab} = 0$. The previous statement allows to assess that the ions dominate the inertial force (when neutrals are negligible or uncoupled) in the framework of MHD description, the electrons follow in quasi equilibrium with a negligible contribution to the inertia and insures local neutrality of the plasma along the flow governed by the Lorentz force. Thus assuming the same velocity for both species and summing the momentum equations of the species, one obtains the MHD equation of motions:

$$\frac{\partial}{\partial t}\rho u_i + \nabla_j (\rho u_i u_i + P_{ij}) = f_i + \rho g_i \; , \tag{11}$$

where $f = J \times B$ is the Lorentz force density, that can be expressed under different forms as will be seen later on. There is no contribution of the electric field because of the local neutrality. The pressure tensor is the sum of all the partial pressure contributions. Its non-diagonal part contents the viscous effect and the tensor is usually decomposed into a scalar part and viscous stress tensor, as follows:

$$P_{ij} = P\delta_{ij} - \tau_{ij} \text{ with } \tau_{ij} = \eta V_{ij} + \zeta\delta_{ij}\text{div}\boldsymbol{u} , \qquad (12)$$

the traceless tensor V_{ij} being the shear tensor : $V_{ij} \equiv \nabla_i u_j + \nabla_j u_i - \frac{2}{3}\delta_{ij}\text{div}\boldsymbol{u}$.

When the medium is weakly ionized, the neutral component is associated to the dynamics and dominates the inertia when the mean free path of the ions in the neutral gas is much shorter than the typical length ℓ_0.

Energy Transport

For a full description an energy transport equation is needed. The summation of the equations of energy transport of each species leads to the equation that governs the transport of the total kinetic energy density $W_{kin} = \frac{1}{2}\rho u^2 + W_{th}$, where W_{th} is the internal energy density:

$$\frac{\partial}{\partial t}W_{kin} + \text{div}(\boldsymbol{u}(\frac{1}{2}\rho u^2 + W_{th} + P) + \boldsymbol{q} - \boldsymbol{u}\cdot\boldsymbol{\tau}) = \boldsymbol{J}\cdot\boldsymbol{E} + \rho\boldsymbol{u}\cdot\boldsymbol{g} . \qquad (13)$$

In collisional plasmas, the heat flux $\boldsymbol{q}$ is proportional to the temperature gradient; this is not the case in collisionless plasmas where Landau effects and microturbulence play a major role. Besides, the transport of the electromagnetic energy of density W_{em}, that involves the Poynting flux $\boldsymbol{S}$, is such that:

$$\frac{\partial}{\partial t}W_{em} + \text{div}\,\boldsymbol{S} = -\boldsymbol{J}\cdot\boldsymbol{E} . \qquad (14)$$

In some cases, an equation of state (i.e. $P(\rho)$) is enough for a fair description.

3 Ohm's Law

In a hot plasma, the validation of Ohm's law is the most tricky task. This law is directly derived from the equation of motion of the electrons after having neglected the inertial force, which is justified when the dynamical time scale is longer than the characteristic time of the electron dynamics (ω_{pe}^{-1}, the inverse of the electron plasma pulsation); it is generally sufficient that the length scale of the phenomenon is much larger than the inertial length of the electrons $\delta_e \equiv c/\omega_{pe}$. Indeed this quasi stationary state is such that

$$\nabla P_e = n_e q_e(\boldsymbol{E} + \boldsymbol{u}_e \times \boldsymbol{B}) - n_e m_e \bar{\nu}_{ei}(\boldsymbol{u}_e - \boldsymbol{u}_i) ; \qquad (15)$$

which allows to write an equation relatively close to the Ohm's law by assuming local neutrality and by introducing the resistivity $\eta \equiv \frac{m_e\bar{\nu}_{ei}}{n_e q_e^2}$:

$$\boldsymbol{E} + \boldsymbol{u}_e \times \boldsymbol{B} = \eta\boldsymbol{J} + \frac{\nabla P_e}{n_e q_e} . \qquad (16)$$

However the equation is a little more complicate if one inserts the fluid velocity instead of the electron fluid velocity. This makes to appear a supplementary contribution, abusively called "Hall effect", $\boldsymbol{J} \times \boldsymbol{B}/n_e q_e$:

$$\boldsymbol{E} + \boldsymbol{u} \times \boldsymbol{B} = \eta \boldsymbol{J} + \frac{\nabla P_e}{n_e q_e} + \frac{\boldsymbol{J} \times \boldsymbol{B}}{n_i q_i} \ . \tag{17}$$

This equation called "generalized Ohm's law" differs from the standard Ohm's law:

$$\boldsymbol{E} + \boldsymbol{u} \times \boldsymbol{B} = \eta \boldsymbol{J} \ . \tag{18}$$

It differs by two extra terms whose importance can be estimated by comparing them to the electro-motive field that is of order $V_A B_0$. In fact the importance of the "Hall term" has already been measured by the importance of the relative velocity between electrons and ions, which leads to

$$\frac{|\boldsymbol{J} \times \boldsymbol{B}/n_i q_i|}{V_A B_0} \sim \frac{r_0}{\ell_0} \ . \tag{19}$$

The pressure term marks the bi-fluids character of the plasma and

$$\frac{|\nabla P_e/n_e q_e|}{V_A B_0} \sim \frac{\delta n}{n} \frac{T_e}{q_e B_0 V_A \ell_0} \sim \beta \frac{\delta n}{n} \frac{r_0}{\ell_0} \ . \tag{20}$$

The parameter β is defined as the ratio of the kinetic pressure over the magnetic pressure; it plays an important role in MHD as will be seen later on.

3.1 Back to Neutrality Assumption

The approximation of local neutrality in MHD regime can be examined more carefully. In that purpose, the relative deviation $\rho_{el}/n_e q_e$ can be examined from Ohm's law by calculating the divergence of the electric field, neglecting the two extra terms of order r_0/ℓ_0 mentioned previously.

$$\frac{\rho_{el}}{n_e q_e} = \frac{\varepsilon_0}{n_e q_e}(\text{div}(\eta \boldsymbol{J}) - \text{div}(\boldsymbol{u} \times \boldsymbol{B})) \ . \tag{21}$$

The term involving the current describes the relaxation of the electric charge, since $\text{div}\boldsymbol{J} = -\frac{\partial \rho_{el}}{\partial t}$. Only the term in $\text{div}(\boldsymbol{u} \times \boldsymbol{B})$ could be at the origin of a charge separation. However, when the divergence does not vanish, its contribution is at most of the order given by

$$\frac{\rho_{el}}{n_e q_e} \sim \frac{\varepsilon_0}{n_e q_e} \frac{u_0 B_0}{\ell_0} \sim \frac{u_0 V_A}{c^2} \frac{r_0}{\ell_0} \ . \tag{22}$$

It is worth noticing that significant deviations can rise in relativistic regime; this is the case of the Pulsar magnetosphere, which has a very fast rotation with an extremely intense magnetic field and generates the so-called Goldreich-Julian charge.

As previously stressed, the validation of MHD description for a mixture of electrons and ions essentially reduces to the condition that the scale ℓ_0 be much larger than the typical ion Larmor radius r_0 that practically coincides with the inertial length of the ions δ_i. When the neutral population is coupled to the plasma through ion-neutral collisions because ℓ_0 is larger than the mean free path, the so-called ambipolar diffusion takes place.

3.2 Ambipolar Diffusion

In some important astrophysical applications, the neutral fluid is the dominant component of the inertia and is coupled to the dynamics. This is the case in the interstellar medium apart from the HII regions, in envelopes of young or giant stars. The MHD description is under collisional conditions and a global description involving the neutral component is valid at scales larger than the mean free path of the ions in the neutral fluid. The ion fluid being more tenuous than the neutral fluid, the ion inertia is negligible, the Alfvén velocity is a function of the neutral mass density and Ohm's law is properly modified.

The plasma having a negligible inertia is thus simply described by an dynamical equilibrium condition:

$$\nabla(P_e + P_i) = \boldsymbol{J} \times \boldsymbol{B} - n_i m_i \bar{\nu}_{in}(\boldsymbol{u}_i - \boldsymbol{u}_n) \tag{23}$$

Ohm's law is modified as follows:

$$\boldsymbol{E} + \boldsymbol{u} \times \boldsymbol{B} = \eta \boldsymbol{J} + \frac{\boldsymbol{B} \times (\boldsymbol{J} \times \boldsymbol{B})}{n_i m_i \bar{\nu}_{in}} , \tag{24}$$

the second term in the left hand side is like an anisotropic resistivity term acting only on the perpendicular current: $\eta_\perp^{amb} = \mu_0 \frac{V_A^2}{\bar{\nu}_{ni}}$. The Alfvén velocity essentially depends on the neutral mass density as long as they are coupled to the ions at the considered scale (i.e. $\ell_0 \gg \bar{\ell}_{in}$).

4 Ideal MHD

The magnetic Reynolds number $\mathcal{R}_m$ is introduced in order to measure the relative importance of the resistive dissipation effect in Ohm's law. Indeed comparing $\boldsymbol{u} \times \boldsymbol{B}$ with $\eta \boldsymbol{J}$, the ratio of the corresponding fiducial quantities defines the magnetic Reynolds number:

$$\mathcal{R}_m \equiv \frac{u_0 \ell_0}{\eta/\mu_0} . \tag{25}$$

When this number is large, the resistive dissipation term can be neglected at first approximation. Moreover if the usual viscous Reynolds number is large also, then dynamics can described without any dissipation effect; this is what one calls "ideal MHD". In general, there always exist some regions of the physical space where that approximation is not valid.

In ideal MHD, the electric field, orthogonal to u and B, has just a compensation role with respect to the motion of the fluid in a transverse direction to the magnetic field; this is equivalent to state that the transverse motion is a general drift of all the components of the fluid at the velocity $E \times B/B^2$.

In standard MHD, a parallel component of the electric field can rise only if the resistivity is not neglected. However in a more general ideal MHD, i.e. when one takes into account the contribution of the electronic pressure in the generalized Ohm's law, a component $E_\parallel$ can develop in order to compensate an electronic pressure gradient along the field lines; which is an effect of order r_0/ℓ_0, as previously seen. This refinement is useful when kinetic corrections, such as Landau effects, are relevant. These are small scale corrections, but often the correct mechanisms for exciting or damping MHD waves.

4.1 Magnetic Diffusion versus "Frozen in Condition"

When the magnetic Reynolds number is weak, resistivity is an important effect and the magnetic field lines diffuse across the plasma. Indeed, eliminating the electric field between the Maxwell-Faraday law ($\frac{\partial B}{\partial t} + \text{rot } E = 0$) and Ohm's law, one gets the following evolution equation (called induction equation):

$$\frac{\partial B}{\partial t} = \text{rot}(u \times B) - \text{rot}(\nu_m \text{rot} B) \ . \tag{26}$$

When $\mathcal{R}_m$ is weak enough and the resistivity is uniform, the equation reduces to a diffusion equation:

$$\frac{\partial B}{\partial t} = \nu_m \Delta B \ . \tag{27}$$

Thus the field tends to become homogeneous in the plasma, because of the dissipation of the electric current, and $\nu_m = \eta/\mu_0$ corresponds to a diffusion coefficient, called magnetic diffusivity.

In ideal MHD, the evolution of the magnetic field is intimately linked with the motions of matter since the field is "frozen" in the plasma flow. The "frozen in" phenomenon stems from two important theorems established by Alfvén.

Theorem 1. *In ideal MHD, the magnetic field lines and more precisely the field B/ρ are transported by the flow, namely: the transformation generated by the flow maps the field B/ρ and its lines at time t_0 with the field B/ρ and its lines at time t.*

What is the meaning of the statement "a vector field is transported by the flow"?

The flow generates a mapping $\phi_{t't}$ of the physical space onto itself which links any point M (x) at every time t with a point M' (x') at time t': $x' = \phi_{t't}(x)$. That differentiable mapping is obtained by integrating the differential system $\dot{y}_i = u_i(y, \tau)$ with the condition $y(t) = x$; which gives the Lagrangian trajectories. Consider a vector field $V(x, t)$. In one hand, on M' at time t', one a priori finds $V(x', t')$, as a definition of a vector field. In other hand, on can also built on M' at time t', a vector $V'(x', t')$ obtained by the flow transportation through the following procedure. On every point M at t, V generates a direction along which one can define an infinitesimal displacement $dl = \varepsilon V$. The mapping of this small displacement by the flow on M/ at t/ is a small displacement dl' such that

$$dl'^i = a^i_j(t', t)dl^j \text{ with } a^i_j(t', t) = \frac{\partial x'^i}{\partial x^j} . \tag{28}$$

For $t' = t + \delta t$, this jacobian matrix can be expanded at first order as

$$a^i_j = \delta^i_j + \delta t \frac{\partial u^i}{\partial x^j} + \mathcal{O}(\delta t^2) . \tag{29}$$

That transformation $\phi^*_{t',t}$ allows to built the field transported by the flow by defining $V' = \varepsilon dl'$ and $V'^i = a(t', t)^i_j V^j$.

If V and V' coincides on every point at every time, one says that the vector field is transported by the flow. In a general way one can evaluate the deviation between these two field by the means of a special derivation operation:

$$\lim_{\delta t \to 0} \frac{V(\phi_{t',t}(x), t') - \phi^*_{t't} V(x, t)}{\delta t} = \frac{\partial V}{\partial t} + \mathcal{L}_u \cdot V , \tag{30}$$

where the second term is, by definition, the Lie derivative of the vector field V with respect to the flow field u:

$$\mathcal{L}_u \cdot V \equiv \lim_{\delta t \to 0} \frac{V(\phi_{t',t}(x), t) - \phi^*_{t',t} V(x, t)}{\delta t} \tag{31}$$

Knowing the expansion of $\phi^*_{t',t}$ at first order and

$$V^i(\phi_{t',t}(x), t) = V^i(x, t) + \delta t\, u^j \frac{\partial V^i}{\partial x^j} + \mathcal{O}(\delta t^2) , \tag{32}$$

one obtains the expression of the Lie derivative of a vector field:

$$\mathcal{L}_u \cdot V \mid^i = u^j \frac{\partial V^i}{\partial x^j} - \frac{\partial u^i}{\partial x^j} V^j \tag{33}$$

That derivation of a vector field is also written as $[\boldsymbol{u}, \boldsymbol{V}]$; the Lie bracket is bilinear, anticommutative, satisfies the Leibnitz rule and Jacobi identity; this operation is quite rich in algebraic properties and helpful for MHD calculation. A field transported by a flow is thus characterized by the following equation:

$$\frac{\partial \boldsymbol{V}}{\partial t} + [\boldsymbol{u}, \boldsymbol{V}] = 0 . \tag{34}$$

In ideal MHD $\boldsymbol{B}/\rho$ is a vector field transported by the flow and, consequently, the field lines are invariant through the transformations generated by the flow. Indeed the induction equation together with $\rho \, \mathrm{div}\, \boldsymbol{u} = -D\rho/Dt$ leads to the following equation, after having developed $\mathrm{rot}(\boldsymbol{u} \times \boldsymbol{B})$:

$$\frac{\partial}{\partial t}\boldsymbol{B} + \boldsymbol{u} \cdot \nabla \boldsymbol{B} - \frac{D\rho}{\rho Dt}\boldsymbol{B} = \boldsymbol{B} \cdot \nabla \boldsymbol{u} . \tag{35}$$

Finally one gets:

$$\frac{D}{Dt}\frac{\boldsymbol{B}}{\rho} = \frac{\boldsymbol{B}}{\rho} \cdot \nabla \boldsymbol{u} \;\Leftrightarrow\; \frac{\partial}{\partial t}\frac{\boldsymbol{B}}{\rho} + [\boldsymbol{u}, \frac{\boldsymbol{B}}{\rho}] = 0 . \tag{36}$$

Any kind of magnetic surfaces can be generated at will. Such a construction has some relevance only if some symmetry suggests the construction. For instance, in an axially symmetric flow, one naturally defines magnetic surfaces preserved by rotations around the axis, each surface is labeled by a definite value of the magnetic flux across an horizontal disk of radius r and centered on the axis at the point of coordinate z. In ideal MHD, the topological genus of the magnetic surfaces is preserved by the flow. For example, if at a time t_0 the surfaces are isomorphic to a sphere, or a torus, or a torus with several holes, they remain so at any further time. However in Nature or in laboratory experiments, topological changes are observed: the magnetic surfaces can be teared and switch from some topology to another one. The tearing sites involve a dynamics which is not governed by ideal MHD; they are sites of "reconnections" of field lines. The reconnection phenomenon is often triggered by a "tearing" instability.

Theorem 2. *In ideal MHD, the magnetic flux across a surface delimited by a closed contour is constant when the contour is deformed by the flow transformation.*

Indeed, during a duration δt the contour γ is transformed into γ' (see Fig. 1); every point of γ is mapped to a point of γ' obtained by $\phi_{t',t}$, $x'^i = x^i + u^i \delta t + \mathcal{O}(\delta t^2)$.

$$\Phi' = \oint_{\gamma'} \boldsymbol{A}' \cdot d\boldsymbol{l}' . \tag{37}$$

The infinitesimal variation $\Phi' - \Phi$ is equal to the difference between the variation $\delta \Phi_1$ due to the modification of the field and the outflux $\delta \Phi_2$ across the band generated by the displacement:

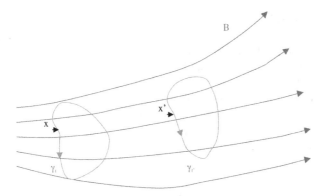

Fig. 1. The oriented contour γ_t at time t is transformed into a contour γ_t' at time t' by the flow. These contours are the boundary of an oriented surface and the flux of the magnetic field across the surface is preserved

$$\delta\Phi_1 = \delta t \oint_\gamma \frac{\partial A}{\partial t} \cdot dl \ , \tag{38}$$

$$\delta\Phi_2 = \oint_\gamma B \cdot (dl \times u\delta t) = \delta t \oint_\gamma (u \times B) \cdot dl \ . \tag{39}$$

Now the variation $\delta\Phi_1$ can be rewritten by inserting the ideal Ohm's law:

$$\delta\Phi_1 = -\delta t \oint_\gamma E \cdot dl = \delta t \oint_\gamma (u \times B) \cdot dl \ . \tag{40}$$

The difference $\delta\Phi_1 - \delta\Phi_2 = 0$ and thus $\Phi' = \Phi$.

5 Confinement and Transverse Diffusion of Matter

5.1 Pressure and Magnetic Tension

The Lorentz force density $f \equiv J \times B$ can be rewritten in another form, illuminating for some theoretical developments. Indeed,

$$f = \frac{1}{\mu_0}\mathrm{rot}B \times B = -\nabla\frac{B^2}{2\mu_0} + \frac{1}{\mu_0}B \cdot \nabla B \ . \tag{41}$$

That force density, which is orthogonal to the field, is thus split into a magnetic pressure contribution that completes the kinetic pressure and a tension contribution. The fact that the second term describes a tension effect can be seen as follows. The term $B \cdot \nabla B$ can also be split into a component that pulls along the line towards the increasing intensity, $\frac{\partial}{\partial s}B^2/2$, but is exactly canceled by the parallel component of the pressure gradient, and a transverse component that pushes towards the direction of the curvature center and tends to make the field line straight, $B^2 \frac{\partial e_b}{\partial s} = -B^2 n/\mathcal{R}$, s is the curviligne coordinate along the considered field line, e_b is the unit vector tangent to the field line, n is the unit normal vector and $\mathcal{R}$ the curvature radius such that $\frac{\partial e_b}{\partial s} = -\frac{n}{\mathcal{R}}$. Thus finally the action is two fold: a transverse pressure gradient and a tension towards the center of curvature.

The picture of piano rods is illuminating to understand ideal MHD. Indeed, when the frozen in effect and the tension effect are combined, the field lines behave as piano rods, having some rigidity and reacting to curvature with a restoring force.

Knowing that div $\boldsymbol{B} = 0$, the Lorentz force density can also be written under the form of divergence of a magnetic stress tensor:

$$f_i = -\frac{1}{\mu_0} \nabla_j (\delta_{ij} \frac{1}{2} B^2 - B_i B_j) \,. \tag{42}$$

The scalar pressure P^* is thus the sum of the kinetic and the magnetic pressures, and the stress tensor τ_{ij}^* is composed of the sum of the kinetic stress and the magnetic stress $B_i B_j / \mu_0$.

An MHD equilibrium is described by

$$\nabla P = \boldsymbol{J} \times \boldsymbol{B} + \rho \boldsymbol{g}$$

An important parameter is introduced in plasma MHD: $\beta \equiv \frac{P}{P_m}$ where $P_m \equiv B^2/2\mu_0$, the magnetic pressure. When it is very small in an equilibrium without gravity, a "force free" approximation can be assumed, namely $\boldsymbol{J} \times \boldsymbol{B} = 0$, and thus $\boldsymbol{J} = \lambda \boldsymbol{B}$. Such equilibrium is conceived as a balance between the magnetic pressure and the tension effect. An equilibrium without any current can also encountered, it is called a "current free" equilibrium; such configuration could be favored in jets.

Complement: Some More About the Energy Transport

The Poynting vector field $\boldsymbol{S} = \frac{\boldsymbol{E} \times \boldsymbol{B}}{\mu_0}$ can be rewritten in MHD by inserting $\boldsymbol{E} = -\boldsymbol{u} \times \boldsymbol{B} + \boldsymbol{E}^*$ where $\boldsymbol{E}^*$ represents any difference to ideal MHD. Then

$$S_i = -\frac{1}{\mu_0}(B^2 \delta_{ij} - B_i B_j)u_j + S_i^* \,,$$

with $\boldsymbol{S}^* = \boldsymbol{E}^* \times \boldsymbol{B}/\mu_0$. The equation of transport of the total energy, kinetic + magnetic, reveals two modified scalar quantities,

$$W^* = W_{th} + W_m \text{ and } P^* = P + P_m$$

and a modified stress tensor

$$\tau_{ij}^* = \tau_{ij} + \frac{1}{\mu_0} B_i B_j \ .$$

It is worth noticing that when $\boldsymbol{E}^* = \eta \boldsymbol{J}$, then $\mathrm{div}(\boldsymbol{E}^* \times \boldsymbol{B}/\mu_0) = -\eta \boldsymbol{J}^2 + \boldsymbol{B} \cdot \mathrm{rot}(\nu_m \boldsymbol{J})$; this second term must not be forgotten in this formulation.

Confinement of a Plasma Column or a Cylindrical Jet

Consider the radial equilibrium of a plasma column along z-axis:

$$\frac{\partial P}{\partial r} = J_\phi B_z - J_z B_\phi$$

with $J_\phi = -\frac{1}{\mu_0} \frac{\partial B_z}{\partial r}$ and $J_z = \frac{1}{\mu_0 r} \frac{\partial r B_\phi}{\partial r}$; which leads to

$$\frac{\partial}{\partial r}(P + \frac{B_z^2}{2\mu_0}) = -\frac{1}{2\mu_0 r^2} \frac{\partial}{\partial r}(r B_\phi)^2 \tag{43}$$

Ampère theorem relates $r B_\phi$ with the current intensity across the horizontal disk of radius r centered on the axis at z: $r B_\phi = \mu_0 I(r)$. Thus one finds that

$$I^2(r) = -\frac{8\pi}{\mu_0} \int_0^r r'^2 \frac{\partial \tilde{P}}{\partial r'} dr' = \frac{16\pi}{\mu_0} \int_0^r r' \tilde{P}(r') dr' - \frac{8\pi}{\mu_0} r^2 \tilde{P} \ , \tag{44}$$

where $\tilde{P} \equiv P + B_z^2/2\mu_0$. As $r \to \infty$, one expects that $r^2 \tilde{P} \to 0$ and that the integral $\int_0^\infty r' \tilde{P}(r') dr'$ converges, in order to get a realization of the plasma confinement by the magnetic tension. When it converges, this latter integral allows to define a characteristic radius a of the column, by stating that it equals to $a^2 \tilde{P}(0)$. The total electric current thus converges and

$$I(\infty) = (\frac{16\pi}{\mu_0} a^2 \tilde{P})^{1/2} \ ; \tag{45}$$

conversely, given a total current, one derives the radius of the plasma column:

$$a = \frac{I(\infty)}{(16\pi/\mu_0 \tilde{P}(0))^{1/2}} \ . \tag{46}$$

Normal Diffusion of Matter

The plasma column, as considered above, seems to be perfectly confined; this is the case in ideal MHD; but a small resistivity is sufficient to allow the matter to diffuse across the magnetic surfaces. Indeed Ohm's law applied for this configuration leads to the following relations: $E_r + u_\phi B_z - u_z B_\phi = 0$, $-u_r B_z = \eta J_\phi$ and $u_r B_\phi = \eta J_z$. Thus there are three possible motions, namely, rotation motion associated to a difference of potential between the magnetic surfaces, together with an arbitrary flow along the axis and a radial diffusion motion allowed by the resistivity effect:

$$u_r = -\eta \frac{J_\phi B_z - J_z B_\phi}{B_\phi^2 + B_z^2} = -\frac{\eta}{B^2} \frac{\partial P}{\partial r} . \tag{47}$$

Inserting the radial motion in the continuity equation (this is the same for both species of particles), one finds the diffusion equation of matter:

$$\frac{\partial}{\partial t} n = \frac{1}{r} \frac{\partial}{\partial r} (r D_\perp \frac{\partial}{\partial r} n) , \tag{48}$$

where the normal diffusion coefficient, of collisional nature, is thus:

$$D_\perp = \eta \frac{P}{B^2} = \frac{\beta}{2} \nu_m . \tag{49}$$

These results holds for both a static column and a cylindrical jet.

The typical confinement time is then deduced from the diffusion coefficient:

$$\tau = \frac{a^2}{2D_\perp} .$$

This time is an important quantity for nuclear fusion experiment in magnetic reactors like Tokomaks. Indeed, when the temperature of a Deuterium-Tritium plasma has reached a value around 100 keV, by an additional heating process (additional because Joule heating has an efficiency that decreases with temperature), the nuclear fusion reactions starts:

$$D + T \rightarrow He^4 + n ;$$

The neutron carries an energy of $14,1\,MeV$ and the alpha particle $3,5\,MeV$. The thermonuclear reaction generates a power density $Q = n_D n_T \epsilon_0 < \sigma v_* >$, where ϵ_0 is the energy carried by the alpha particle (the neutron energy is lost by the plasma). For a given value of $n_D + n_T$, the product is maximum for $n_D = n_T = n$. Some amount of heat is lost by particle diffusion across the magnetic surfaces at a rate depending on the confinement time τ: $Q_{loss} = -3nT/\tau$. The Lawson criteria gives the condition for this lost to be compensated by the production of nuclear energy. For $T = 100\,keV$, one finds

$$n\tau > 10^{14} cm^{-3}.s . \tag{50}$$

This criteria is more severe when one takes into account the various energy conversion yields and the radiative lost. The tendency is to built bigger and bigger machines in order to increase the confinement time.

6 MHD Waves in Homogeneous Plasma

In this section, the calculation of the MHD modes is presented in the simplest case where the mean field is homogeneous $B_0 = B_0 e_z$, the unperturbed plasma is homogeneous: $\rho = \rho_0 = constant$ and without boundary.

6.1 The Alfvén Wave

The competition between inertia and tension gives rise to an incompressible transverse wave, called Alfvén wave. The tension effect that is described by the force density field $\boldsymbol{f}_T = \frac{1}{\mu_0}\boldsymbol{B}\cdot\nabla\boldsymbol{B}$ can be linearized for small perturbation and reads

$$\boldsymbol{f}_T = \frac{1}{\mu_0}B_0\frac{\partial}{\partial z}\delta\boldsymbol{B} \ ,$$

which is divergence free. Since an incompressible perturbation is considered, $\delta\rho = 0$ and div $\boldsymbol{u} = 0$. The total pressure perturbation vanishes. This stems from the fact that its laplacian necessarily vanishes in an incompressible flow, since the inertial density force is divergence free as well; moreover, the plasma is supposed homogeneous and unlimited, therefore not only the Laplacian but also the total pressure fluctuation vanishes. Since $\delta P^* = C_s^2\delta\rho + B_0\delta B_\|/\mu_0$, and $\delta\rho = 0$, also $\delta B_\| = 0$.

The linearized equation of motions is thus:

$$\rho_0\frac{\partial}{\partial t}\boldsymbol{u} = \boldsymbol{f}_T = \frac{1}{\mu_0}B_0\frac{\partial}{\partial z}\delta\boldsymbol{B} \tag{51}$$

and the induction equation together with the incompressibility condition reads

$$\frac{\partial}{\partial t}\delta\boldsymbol{B} + [\boldsymbol{u},\boldsymbol{B}] = 0 \tag{52}$$

whose linearization in a uniform mean field reduces to

$$\frac{\partial}{\partial t}\delta\boldsymbol{B} - B_0\frac{\partial}{\partial z}\boldsymbol{u} = 0 \ . \tag{53}$$

Rederiving the equation according to time and inserting the first one, one obtains the equation of propagation of the Alfvén waves:

$$\frac{\partial^2}{\partial t^2}\delta\boldsymbol{B} - V_A^2\frac{\partial^2}{\partial z^2}\delta\boldsymbol{B} = \boldsymbol{0} \ . \tag{54}$$

This is a typical transverse wave with $\{\boldsymbol{E},\delta\boldsymbol{B},\boldsymbol{B}_0\}$ forming a direct orthogonal trihedron. The frequency is such that $\omega = k_\|V_A$ and the group velocity points along the direction of the mean field and is nothing but the Alfvén velocity.

6.2 The Magneto-sonic Waves

Because the total pressure depends on the mass density and on the parallel component of the magnetic field perturbations, waves of sonic type can be excited involving a coupling between $\delta\rho$ and $\delta B_{\parallel}$. A simple derivation of the system of coupled equations for $\delta\rho$ et $\delta B_{\parallel}$ can be obtained as follows.

First one linearizes the mass conservation equation:

$$\frac{\partial}{\partial t}\frac{\delta\rho}{\rho_0} + \operatorname{div}\boldsymbol{u} = 0 \tag{55}$$

Since the tension contribution is divergence free ($\operatorname{div}\boldsymbol{f}_T = 0$), one eliminates its contribution by taking the divergence of the two parts of the linearized equation of motions:

$$\frac{\partial}{\partial t}\operatorname{div}\boldsymbol{u} + \frac{1}{\rho_0}\Delta\delta P_* = 0 \ . \tag{56}$$

Which leads to

$$\frac{\partial^2}{\partial t^2}\frac{\delta\rho}{\rho_0} - C_s^2\Delta\frac{\delta\rho}{\rho_0} = V_A^2\Delta\frac{\delta B_{\parallel}}{B_0} \tag{57}$$

The parallel component of the induction equation reads:

$$\frac{\partial}{\partial t}\frac{\delta B_{\parallel}}{B_0} = \frac{\partial}{\partial z}u_z - \operatorname{div}\boldsymbol{u} \tag{58}$$

One expresses $\operatorname{div}\boldsymbol{u}$ as a function of $\delta\rho/\rho_0$, as for u_z, it is easily determined by the projection of the motion equation along the mean field which is devoid of any Lorentz force contribution:

$$\frac{\partial}{\partial t}u_z + C_s^2\frac{\partial}{\partial z}\frac{\delta\rho}{\rho_0} = 0 \ . \tag{59}$$

Taking the time derivative of the evolution equation for $\delta B_{\parallel}$, one finds thus

$$\frac{\partial^2}{\partial t^2}\frac{\delta B_{\parallel}}{B_0} = -(C_s^2\frac{\partial^2}{\partial z^2}\frac{\delta\rho}{\rho_0} - \frac{\partial^2}{\partial t^2}\frac{\delta\rho}{\rho_0}) \ . \tag{60}$$

Therefore the system of equations governing the propagation of the magnetosonic waves is obtained:

$$\frac{\partial^2}{\partial t^2}\frac{\delta\rho}{\rho_0} - C_s^2\Delta\frac{\delta\rho}{\rho_0} - V_A^2\Delta\frac{\delta B_{\parallel}}{B_0} = 0 \tag{61}$$

$$\frac{\partial^2}{\partial t^2}\frac{\delta B_{\parallel}}{B_0} - V_A^2\Delta\frac{\delta B_{\parallel}}{B_0} - C_s^2\Delta_\perp\frac{\delta\rho}{\rho_0} = 0 \ . \tag{62}$$

The dispersion relation is deduced by stating that the following determinant vanishes:

$$\begin{vmatrix} \omega^2 - k^2 C_s^2 & -k^2 V_A^2 \\ -k_\perp^2 C_s^2 & \omega^2 - k^2 V_A^2 \end{vmatrix}$$

which gives

$$\omega^4 - k^2(C_s^2 + V_A^2)\omega^2 + k_{\parallel}^2 k^2 C_s^2 V_A^2 = 0 \ . \tag{63}$$

The dispersion equation has two couples of roots, each couple containing a forward and a backward wave of opposite velocity. These correspond to the slow and the fast magneto-sonic modes of phase velocity $V = \omega/k$. The propagation angle being θ, $k_\parallel = k\cos\theta$. For $\theta = 0$, the slow mode has a velocity equals to the sound velocity $V_s = C_s$ and the fast mode has a velocity V_f that coincides with the velocity of the Alfvén mode; indeed the fast and Alfvén modes degenerate. For almost perpendicular propagation (i.e. $\cos^2\theta \ll 1$), the fast mode reaches its highest velocity:

$$V_f = \sqrt{V_A^2 + C_s^2}$$

and the slow mode velocity tends to vanish:

$$V_s \simeq k_\parallel C_s \frac{V_A}{\sqrt{V_A^2 + C_s^2}} \cdot$$

7 Main Applications and "Special Effects"

Some special effects are attributed to the magnetic field.

- The confinement effect is the simplest one and plays an important role not only in laboratory plasma physics but also in astrophysics, in particular for the collimation of jets.
- There are specific heating processes that are due to the magnetic field, like the "magnetic pumping" effect.
- The "frozen in" effect, previously explained, is very important in astrophysics and very often invoked in the phenomenology of objects formation or in the dynamics of astrophysical media.
- Because the magnetic field generates some internal pressure, it participates to the Archimede thrust. And because of the frozen in condition, magnetic tubes, that are lighter than equivalent unmagnetized plasma volumes, are pushed upwards. This is the magnetic buoyancy effect responsible for the emergence of flux tubes from the photosphere into the solar corona.
- Because the magnetic field generates a specific stress, as previously seen, it can efficiently produce angular momentum transfer. This is a particularly important process for the accretion-ejection phenomenon.
- Dynamo action is one of the most famous MHD effect (see the excellent introduction by Moffat [4]) that is still under intensive investigations in MHD theory development, in numerical simulations and also in experiments that are currently being built.
- MHD has specific waves, solitons, and shocks that are different in their nature and properties from the corresponding hydrodynamical concepts.
- MHD has specific instabilities governed by Lorentz force that are not encountered in Hydrodynamics (see [2, 3]).

Besides technological applications, the main applications of MHD are encountered either in laboratory plasma physics, such as nuclear fusion experiments, or in close astrophysics, such as the physics of planetary magnetospheres, of the Sun and the solar wind, or in remote astrophysics, such as the physics of magnetized stars, white dwarfs, neutron stars and pulsars, black hole environments, jets, the physics of particle accelerating shocks etc.

8 Magnetic Reconnections

8.1 General Considerations

The magnetic reconnection phenomenon is a very important in Tokomak physics, space physics and astrophysics. It is a dissipation process that occurs at very small scale in a very small region, but that manifests as a powerful generation of heat and energetic particles together with some acceleration of the plasma. The energy input in a reconnection zone is a flux of magnetic energy. It takes place in regions where the direction of the field lines changes suddenly, i.e. over a short scale, as for instance in the magneto-tail of the Earth where the field lines are in opposite directions as one crosses the "neutral sheet".

It is useful to first consider a reconnection site with a geometrical viewpoint. The reconnection site is surrounded by a plasma described by ideal MHD with a good approximation. At least locally, one can generally define magnetic surfaces quite naturally thanks to some symmetry. As previously seen, the magnetic surfaces keep their topological genus in ideal MHD because of the "frozen in" condition; in particular, they cannot be teared. When the orientation of the field lines changes on a short scale, the ideal MHD condition is violated. A localized strong current generates a strong Lorentz force density between these magnetic surfaces that attracts them towards each other. The magnetic surfaces touch each other and change their topology: "conjugate surfaces" have a hole and weld each other, then allowing field lines to lay on one side of the new welded surface to go through the hole and to lay on the other side of the welded surface with an almost opposite direction (see Fig. 1).

Actually the physics of the 3D reconnection suggested by the geometrical picture is not yet mastered. It has been intensively studied in the 2D approximation where the dependence according to the coordinate along the current sheet is disregarded.

The reconnection sites need to be generated by dynamical effects like those developed by the "tearing" instability which, as its name suggests, leads to tear the magnetic surfaces by violating ideal MHD. This instability will not be presented in this short introduction. However it is often possible to treat a reconnection site in a stationary state even-though the process looks very eruptive like in solar bursts. This is because the life time of the reconnection site is very long compared to the Alfvén time. At the innermost region of a reconnection site, the electric current is concentrated on a sheet of thickness δ much smaller than any other scales such as the sheet width L and the sheet length L_z. In the vicinity of the current sheet the ideal approximation is violated and, for many years, reconnection has been described in the frame work of resistive MHD. The first historical works of that kind [6, 11] introduced the famous Sweet & Parker mechanism; an extension of the model involving slow shocks has been proposed by Petschek [8], which had a great success in the astrophysical community. A more recent approach involving a more refined description than MHD leads to successful results and turns out to be a promising solution of the long standing issue of reconnection (see a recent book devoted to this issue by Biskamp [1]). In this new approach there is a transition zone between ideal MHD and the dissipation sheet governed by "whistler" dynamics that controls the mass and energy fluxes. Therefore the violation of ideal MHD is not due to resistive effects but to kinetic effects.

In this lecture I will briefly present the celebrated Sweet-Parker model and then the whistler model; I have not enough time for presenting the Petchek model (see [9]). These models are all stationary, two-dimensional and display an odd mirror symmetry with respect to the plane $\{y = 0\}$ (see Fig. 2). Because the magnetic field reverses its orientation over a small scale δ a strong current along the z-direction is concentrated on a sheet of thickness δ, the width L along the x-axis is larger and depends on the details of the model.

8.2 The Sweet-Parker Model

The concentration of the current comes from Ampère law:

$$\boldsymbol{J} = \frac{1}{\mu_0}\mathrm{rot}\boldsymbol{B} \implies \boldsymbol{J} = J\boldsymbol{e}_z(1 + \mathcal{O}(\frac{\delta}{L})) \text{ and } J \sim \frac{B_i}{\mu_0\delta} \tag{64}$$

The electric field, $\boldsymbol{E} = E\boldsymbol{e}_z$, is derived from Ohm's law: outside the current sheet (ideal MHD) $E - u_y B_x = 0$, and thus

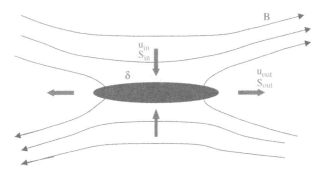

Fig. 2. 2D-reconnection governed by resistive MHD. The size of the current sheet δ is such that the magnetic Reynolds number equals unity. Roughly half of the magnetic energy influx is converted into heat flux, the other half converted into kinetic energy flux, with an outflow at the Alfvén speed

$$E = u_i B_i \tag{65}$$

Inside the current sheet around the stagnation point,

$$E = \eta J \sim \nu_m \frac{B_i}{\delta} \ . \tag{66}$$

Therefore

$$u_i = \frac{E}{B_i} \sim \frac{\nu_m}{\delta} \implies \mathcal{R}_i \sim 1 \ . \tag{67}$$

Inside the current sheet of thickness δ, the magnetic Reynolds number is weak; which means that magnetic diffusion is taking place efficiently and smoothes out the magnetic field irregularities and thus allows *reconnections* of twisted field lines.

Matter Flux

Despite a smooth density variation, the conservation of mass flux implies

$$u_{out} \sim \frac{L}{\delta} u_i \tag{68}$$

Conservation of the Magnetic Flux Outside the Sheet

$$[\boldsymbol{u}, \frac{\boldsymbol{B}}{\rho}] = 0 \tag{69}$$

and $\frac{\partial}{\partial y} \sim \frac{1}{\delta}$ implies

$$u_y \frac{\partial}{\partial y} \frac{B_x}{\rho} - \frac{B_y}{\rho} \frac{\partial}{\partial y} u_x = 0 \ , \tag{70}$$

and a smooth variation of ρ leads to

$$B_{out} \sim \frac{\delta}{L} B_i \tag{71}$$

Motions of the Outflow

The Lorentz force density pushes the flow in the direction of the curvature center

$$f_x = -J_z B_y, > 0 \text{ for } x > 0 \text{ and } < 0 \text{ for } x < 0 \ . \tag{72}$$

Thus

$$f_{out} \sim J B_{out} \sim \frac{B_i B_{out}}{\mu_0 \delta} \sim \frac{B_i^2}{\mu_0 \rho} \tag{73}$$

$$\rho \frac{u_{out}^2}{L} \sim f_{out} \implies u_{out}^2 \sim \frac{B_i^2}{\mu_0 \rho} \tag{74}$$

$$u_{out} \sim V_{Ai} \tag{75}$$

Energy Budget

The sheet is heated by Joules effect:

$$P_J/L_z \sim \eta J^2 L\delta \sim \eta \frac{B_i^2}{\mu_0^2} \frac{L}{\delta} \sim u_i \frac{B_i^2}{\mu_0} L \tag{76}$$

The power dissipated by Joule effect is a sizable fraction of the Poynting influx.

The outflux of kinetic energy is estimated as follows:

$$P_{out}/L_z \sim \frac{1}{2}\rho u_{out}^3 \delta \sim \frac{1}{2}\rho V_{Ai}^2 u_{out}\delta \sim \frac{1}{2}\frac{B_i^2}{\mu_0} u_i L \ . \tag{77}$$

The outflux of kinetic energy is thus a sizable fraction of the Poynting influx as well.

However part of the energy budget is not mastered in the 2D model because particle acceleration takes place along the current sheet. Indeed, because of the electric field that points along z-axis, run away electrons are accelerated and reach an energy equal to the difference of potential along the line:

$$\epsilon_{max} = ZeEL_z \sim Zeu_i B_i L_z \ . \tag{78}$$

Since the length of the line, L_z, is unknown, it is impossible to control how much power goes into particle acceleration. Of course this power is also at maximum a sizable fraction of the Poynting influx.

The Main Problem of the Model

The mass flux and the energy flux vanish when resistivity goes to zero.

$$u_i L \sim u_{out}\delta \sim \delta V_{Ai} \tag{79}$$

and

$$\delta \sim \left(\frac{L\nu_m}{V_{Ai}}\right)^{1/2} \ . \tag{80}$$

This is the main problem for astrophysical plasmas that have a very large magnetic Reynolds number and for which a fast and efficient reconnection process is expected.

8.3 Reconnection Mediated by Whistler Dynamics

The thickness δ of the current sheet, which is also the variation scale of the magnetic field in the y-direction, is often smaller than the minimal scale δ_i (or equivalenty r_0 as seen at the first section) required for the validation of the MHD description. Indeed

$$\frac{\delta}{\delta_i} = \sqrt{\frac{m_e}{m_i}} \frac{c}{u_i} \frac{\nu_{ei}}{\omega_{pe}} , \tag{81}$$

and is almost always much smaller than unity in astrophysical plasmas, in space plasmas and in tokomak plasmas. Therefore a description finer than MHD description is required. In order to properly improve the description, it is important to stress the following points:

- The magnetic field is frozen in the electron flow as long as the dynamics develop at a scale larger then the electron inertial scale defined by $\delta_e \equiv c/\omega_{pe}$.
- The electrons have a negligible inertia at scales larger than δ_e.
- The ions are no longer magnetized when they move at scales smaller than $r_0 \equiv V_A/\omega_{ci}$.
- The ions are very inert at scale smaller than their inertial scale defined by $\delta_i \equiv c/\omega_{pi}$. Now it has been indicated that the inertial length of the ions coincides with r_0, the minimum MHD scale.

The non-collisional reconnection process is elaborated on the following fundations (see Fig. 3). The reconnection sheet is concentrated over a thickness of order δ_e. Between δ_e and δ_i, the magnetic field is frozen in the electron flow, the motions are such that the electrons have a negligible inertia and the ions are almost inert. That intermediate dynamics is typical of the "whistler" wave that can be derived under the conditions $m_e \to 0$ and $m_i \to \infty$. The dynamics is not governed by mechanics anymore, this is pure electrodynamics with a current carried by the electron fluid only:

$$\boldsymbol{J} \simeq -n_e e \boldsymbol{u}_e . \tag{82}$$

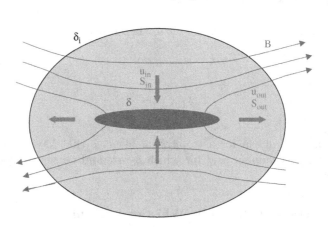

Fig. 3. 2D1/2-reconnection governed by whistler dynamics. Between the smallest MHD scale δ_i and the size of the current sheet $\delta \sim \delta_e$, there is an intermediate region where whistler dynamics takes place and controls the influx and the outflux of mass and energy

At that approximation, the electric field $\boldsymbol{E}+\boldsymbol{u}_e\times\boldsymbol{B}=\boldsymbol{0}$ is nothing but the Hall electric field. The fluxes of matter and energy between the MHD domain and the current sheet are thus governed by what is called the "whistler dynamics".

Like in Sweet-Parker model, $u_iL \sim u_{out}\delta$ and $u_iB_i \sim u_{out}B_{out}$. But the motions are those of the electrons and the speed of the outflow is a kind of Alfvén velocity with the electron mass:

$$u_{out} \sim \frac{B_i}{\sqrt{\mu_0 n_e m_e}} \equiv V_{Ai}^e \ . \tag{83}$$

This result is not derived from a balance between the inertial force and the Lorentz force like Sweet-Parker model; it comes from a purely electrodynamic calculation (see below).

The matter influx is very fast and the Poynting influx is carried by the electrons is very intense, which leads to an efficient and fast reconnection process. The products

$$u_iL \sim u_{out}\delta \sim V_{Ai}^e\delta_e \sim \frac{B_i}{\mu_0 n_e e} \ , \tag{84}$$

are independent of any mass and resistivity and lead to an universal budget free from any kind of dissipation process taking place in the current sheet. Numerical simulations [10] show that $L \sim 10\delta_e$. The Poynting flux entering the reconnection site is much more intense than in the Sweet-Parker model, because the field is transported by the fast electrons motions. The outflux of kinetic energy is a sizable fraction of the Poynting influx:

$$P_{out}/L_z \sim \frac{1}{2}n_e m_e u_{out}^3\delta \sim \frac{1}{2}n_e m_e(V_{Ai}^e)^2 u_{out}\delta \sim \frac{1}{2}\frac{B_i^2}{\mu_0}u_iL \ . \tag{85}$$

Whatever the kind of dissipation taking place in the current sheet, the power dissipated seems to be always a sizable fraction of the Poynting influx as well. Indeed extrapolating the value of the reconnection electric field (which is almost constant at first approximation) $E = u_iB_i$ in the sheet, knowing that there is a current J_z along the neutral sheet, one finds that the power transmitted by the electric field in the sheet is such that $P_E/L_z = J_zE\delta L \sim u_iB_i^2/\mu_0$. A sizable fraction of the Poynting flux is really found and it contributes to heat thermal electrons and to accelerate suprathermal electrons in the sheet.

Calculation of the Outflow

The frozen in condition in the electron flow reads $[\boldsymbol{u}_e, \boldsymbol{B}/\rho] = 0$ and, like in Sweet-Parker, implies

$$u_yB_x \sim B_yu_x \ ,$$

and also

$$u_yB_z \sim B_yu_z \ .$$

An important difference with Sweet-Parker model is that all the fields necessarily have three components. Maxwell-Amp'ere equations provide three other
relations:

$$u_x = \frac{1}{\mu_0 n_0 e} \frac{\partial B_z}{\partial y} \tag{86}$$

$$u_y = \frac{1}{\mu_0 n_0 e} \frac{\partial B_z}{\partial x} \tag{87}$$

$$u_z = \frac{1}{\mu_0 n_0 e} \left(\frac{\partial B_x}{\partial y} - \frac{\partial B_y}{\partial x} \right) . \tag{88}$$

From the ratio of the first and the third components and of the two relations
provided by the induction equation, one derives that

$$\frac{B_z}{B_x} \sim \frac{u_z}{u_x} \sim \frac{B_x}{B_z}$$

and thus $u_z \sim u_x$ et $B_z \sim B_x$.

Therefore the following set of results is obtained:

$$u_{out} \sim \frac{1}{\mu_0 n_0 e} \frac{B_i}{\delta} \tag{89}$$

$$u_i \sim \frac{1}{\mu_0 n_0 e} \frac{B_i}{L} \tag{90}$$

$$u_z \sim \frac{1}{\mu_0 n_0 e} \frac{B_i}{\delta} \tag{91}$$

and also $B_{out} \sim \frac{\delta}{L} B_i$ et $B_z \sim B_i$. That important generation of motions and
field component along the neutral sheet marks a profound difference with the
Sweet-Parker model.

8.4 Perspective

These results have been checked by three types of numerical simulations:
particles in cell, hybrid, Hall MHD (see the excellent paper by Shay et al.
[10]). One of the main efforts in that field is the extension of the results to 3D
situations. The existence of a whistler range governing the fluxes is confirmed.
The differences with the 2D description is the development of instabilities
generated by the electron outflow, which excite plasma modes that are not
taken into account by MHD even by "Hall MHD". Hall MHD is the simplest
extension of MHD that takes into account the Hall effect in the generalized
Ohm's law. Although it does not describe the whole phenomenology of the
reconnection process, the use of the Hall MHD is sufficient to catch the main
aspect of the phenomenology of reconnection, namely the fluxes of matter and
energy, since, as previously seen, these results are independent of the details
of the dissipation process at very small scales. The Hall MHD contains other

modes than MHD modes, in particular the whistler mode that has a frequency proportional to the square of the wave number, (hence its name, because high frequency waves arrive before low frequency waves when a whistler wave packet is generated in the ionosphere, for instance). This is in fact a difficulty for developing numerical codes (look at [5]), because the short scale numerical errors propagate faster than the smoother ones ... Nevertheless it seems that the long standing issue of magnetic reconnection will be mastered on the near future. Which is a good news for theoretical astrophysics.

The non-linear development of the Magneto-Rotational Instability (MRI) in accretion disks depends very much of a correct description of the reconnection process. Another important MHD process that depends also very much on reconnections is the Dynamo action in stars, accretion disks, galaxies, galaxy clusters etc. The progress in the theoretical and numerical investigations of this processes depends on the progress in understanding reconnections.

References

1. Biskamp D., 2000, "Magnetic Reconnection in Plasmas", Cambridge University Press.
2. Freidberg J.P., 1987, Ideal Magnetohydrodynamics. Plenum Press, New York.
3. Goedbloed H., Poedts S., 2004, Principles of MagnetoHydrodynamics, Cambridge University Press.
4. Moffat H.K., 1978, "Magnetic field generation in electrically conducting fluids", Cambridge University Press.
5. O'Sullivan S., Downes T., 2006, MNRAS, 366, 1329.
6. Parker E.N., 1963, Astrophys. J. Suppl., 8, 177.
7. Parker E.N., 1979, "Cosmical Magnetic Fields" Clarendon Press Oxford.
8. Petschek H.E., 1964, "Magnetic field annihilation". in The physics of solar flares, NASA Publ. SP-50, p. 425.
9. Priest E.R., 1984, "Solar Magneto-Hydrodynamics", D. Reidel, Dordrecht.
10. Shay M.A., Drake J.F., Rogers, B.N., Denton R.E., 2001, J. Geophys. Res., 106, 3759.
11. Sweet P.A., 1958, "The neutral point theory of solar flares", in Electromagnetic Phenomena in Cosmical Physics, IAU Symposium 6 (Cambridge University Press).

Theory and Models of Standard Accretion Disks

Caroline E. J. M. L. J. Terquem

Institut d'Astrophysique de Paris, UMR7095 CNRS, Université Pierre et Marie Curie–Paris 6, 98bis boulevard Arago, 75014 Paris
Université Denis Diderot–Paris 7, 2 Place Jussieu, 75251 Paris Cedex 5
Institut Universitaire de France
terquem@iap.fr

1 Introduction

It is an observational fact that at least part of the mass in protostellar disks is accreted onto the central object. In a Keplerian accretion disk, the specific angular momentum increases with radius. Therefore, a particle can be accreted only if its angular momentum is removed or transferred to particles located at larger radii. Whether angular momentum is removed from or redistributed within the disk depends on whether the disk is subject to external or internal

C. E. J. M. L. J. Terquem: *Theory and Models of Standard Accretion Disks*, Lect. Notes Phys.
723, 103–115 (2007)
DOI 10.1007/978-3-540-68035-2_5   © Springer-Verlag Berlin Heidelberg 2007

torques. Possible external torques can either be magnetic (when an outflow is present) or tidal (in binary systems), whereas possible internal torques can either be gravitational (massive disks) or turbulent. These mechanisms have been discussed by Papaloizou & Lin (1995).

During the early stages of disk evolution, when the disk is still embedded (class 0/I object) and has a significant mass compared to the central star, there may exist strong disk winds and bipolar outflows (e.g. [23]) with associated magnetic fields. During this stage, a hydromagnetic disk wind may be an important means of angular momentum removal for the system [5]. Because of the action of magnetic torques, material ejected from the disk is able to carry away significantly more angular momentum than it originated with in the disk. Therefore, even a modest ejection rate can lead to a significant accretion rate through the disk. However, observations indicate that outflows may exist only in the early stages of disk evolution, so that this mechanism cannot account for angular momentum transport during the whole life of accretion disks. In addition, it may affect only the inner parts.

When the mass of the disk is significant compared to that of the star, gravitational instabilities may develop, leading to outward angular momentum transport [14, 15, 20, 21] that results in additional mass growth of the central star. This redistribution of mass may occur on the dynamical timescale (a few orbits) of the outer part of the disk and thus may be quite rapid: on the order of 10^5 yr for a disk radius of 500 AU. The parameter governing the importance of disk self–gravity is the Toomre parameter, $Q \sim M_* H/(M_d r)$, with M_* being the central mass, M_d being the disk mass contained within radius r and H being the disk semi–thickness. Typically $H/r \sim 0.1$ [25] such that the condition for the importance of self–gravity, $Q \sim 1$, gives $M_d \sim 0.1 M_*$. During the period of global gravitational instability, it is reasonable to suppose that the disk mass is quickly redistributed and reduced by accretion onto the central object such that the effects of self–gravity become negligible.

If the disk surrounds a star which is in a pre–main sequence binary system, tidal torques transport angular momentum outward, from the disk rotation to the orbital motion of the binary. However, although tidal effects are important for truncating protostellar disks and for determining their size, it is unlikely that tidally–induced angular momentum transport plays a dominant role in the evolution of protostellar disks (see Terquem 2001 and references therein). In a non self–gravitating disk, the amount of transport provided by tidal waves is probably too small to account for the lifetime of protostellar disks. In addition, tidal effects tend to be localized in the disk outer regions.

When the disk mass is such that self–gravity can be ignored and the jet activity has significantly decreased, turbulent torques may become the most important way of redistributing angular momentum in the disk. Historically, the first angular momentum transport mechanism to be considered was through the action of viscosity [30]. However, in order to result in evolution on astronomically interesting timescales, it is necessary to suppose that an anomalously large viscosity is produced through the action of some sort of turbulence.

In this paper we present the theory and models of turbulent accretion disks. For more detail on disk theory, the reader is referred to the reviews by Papaloizou & Lin (1995) and Balbus & Hawley (1998) and to the book by Frank et al. (2002). Disk models are reviewed by Dullemond et al. (2006).

2 Turbulent Angular Momentum Transport in Disks

The molecular kinematic viscosity is given by:

$$\nu \sim c_s \lambda \,, \tag{1}$$

where c_s is the sound speed and λ is the mean free path of the particles. In a typical protostellar disk, at the distance $r = 1$ astronomical unit (au) from the central star, $\lambda \sim 1$ m and $c_s \sim 10^3$ m s^{-1}. The associated viscous timescale $t_\nu \sim r^2/\nu$ is therefore larger than the age of the Universe!

The realization that molecular transport of angular momentum is so inefficient led the theorists to look for another mechanism of transport in accretion disks. Because Reynolds numbers are so high, it was thought that probably accretion disks would be subject to the same hydrodynamical nonlinear instabilities that shear flows experience in laboratory. The resulting turbulence would then transport angular momentum efficiently. Although today much doubt has been cast on hydrodynamical instabilities in disks, turbulence is still a strong candidate for transport since it has been shown relatively recently that a linear magnetohydrodynamical instability can develop in disks (see below). Therefore, we turn now to turbulent transport, and contrast it with molecular transport. Much of this section is based on Tennekes & Lumley (1972), which the reader is referred to for more details (see also [3]).

We shall restrict ourselves here to the study of incompressible flows, as this simplifies the discussion. For our argument, it is sufficient to take into account only pressure and viscous forces, but in principle any other (external or inertial) force could be added. The equations describing the fluid are the Navier–Stokes equation of motion:

$$\frac{\partial \tilde{v}_i}{\partial t} + \tilde{v}_j \frac{\partial \tilde{v}_i}{\partial x_j} = \frac{1}{\rho} \frac{\partial}{\partial x_j} \tilde{\sigma}_{ij} \,, \tag{2}$$

and the equation of mass conservation:

$$\frac{\partial \tilde{v}_i}{\partial x_i} = 0 \,, \tag{3}$$

where the x_i ($i = 1, 2, 3$) denote the coordinates and we adopt the standard Einstein notation of summation over repeated indices. Here $\mathbf{v}$ is the fluid velocity, ρ is the density of mass and $[\sigma]$ is the viscous stress tensor. The tilde symbol above a variable means that we consider the instantaneous value of the variable at the location x_i and time t. We have:

$$\tilde{\sigma}_{ij} = -\tilde{p}\delta_{ij} + \eta\tilde{s}_{ij} , \qquad (4)$$

where $\tilde{s}_{ij}$ in an incompressible fluid is given by:

$$\tilde{s}_{ij} = \frac{\partial \tilde{v}_i}{\partial x_j} + \frac{\partial \tilde{v}_j}{\partial x_i} . \qquad (5)$$

Here $\tilde{p}$ is the pressure and $\eta = \rho\nu$ is the shear viscosity that we will take constant.

We now use the so–called Reynolds decomposition, in which an instantaneous value is written as the sum of a mean value (denoted by a capital letter) plus a fluctuation (denoted by a small letter):

$$\tilde{v}_i = V_i + v_i . \qquad (6)$$

This fluctuation is characteristic of the turbulence. This decomposition is meaningful only if the timescale on which the fluctuations vary and the evolution timescale of the flow are well separated. The mean values are then taken over a timescale large compared to the turbulence timescale but short compared to that of the flow evolution. As here we are not interested in the long term evolution of the flow, we neglect the derivative with respect to time of the mean values (i.e. we consider a quasi–steady state). To simplify the discussion, we suppose that the average of v_i over time is zero: $< v_i >= 0$.

Equation (3) averaged over time then leads to $\partial V_i/\partial x_i = 0$, i.e. the mean flow is incompressible. Equation (3) thus implies $\partial v_i/\partial x_i = 0$, i.e. the fluctuations are also incompressible.
Using

$$\tilde{\sigma}_{ij} = \Sigma_{ij} + \sigma_{ij} , \qquad (7)$$

with $\langle \sigma_{ij} \rangle = 0$, the equation of motion averaged over time gives:

$$\frac{\partial}{\partial x_j} V_j V_i + \frac{\partial}{\partial x_j} \langle v_j v_i \rangle = \frac{1}{\rho}\frac{\partial}{\partial x_j}\Sigma_{ij} , \qquad (8)$$

where we have used $\partial V_i/\partial t = 0$ and the incompressibility of the mean flow and the fluctuations. The term $< v_j v_i >$ represents the averaged transport of the fluctuations of the momentum by the fluctuations of the velocities. This is the turbulent transport. It is non zero only if the turbulent velocities in the different directions are correlated. It is an experimental fact that this is in general the case for shear turbulence. Equation (8) shows that momentum is transferred between the fluctuations and the mean flow through the term $\partial < v_j v_i > /\partial x_j$.

We can rewrite (8) under the form:

$$\frac{\partial}{\partial x_j} V_j V_i = \frac{1}{\rho} \frac{\partial}{\partial x_j} \left(\Sigma_{ij} + \tau_{ij} \right) , \tag{9}$$

where $\tau_{ij} = -\rho \langle v_j v_i \rangle$ is called the *Reynolds stress tensor*, or turbulent stress. This is the contribution from the fluctuations to the averaged total stress tensor $\Sigma_{ij} + \tau_{ij}$. Note that if we had not supposed $< v_i >= 0$, we would also have had terms like $V_j < v_i >$, representing transport (or advection) of mean momentum by the fluctuations. This is what Lynden–Bell & Kalnajs (1972) call *lorry–transport*, because it is a direct "shipment" by the equilibrium flow. Formally, this term is not part of what we call the turbulent stress however.

As we have no expression for the components of τ_{ij} (six of which are independent), the problem has more unknowns than equations. This is the well–known closure problem of turbulence. Since τ_{ij} appears in the same way as Σ_{ij} in (9), it is tempting to express τ_{ij} by analogy with Σ_{ij}, which depends on the molecular motion. This is the basis for the *mixing length theory*, in which τ_{ij} is written exactly like the tensor deriving from molecular motions using a so–called turbulent viscosity ν_T. By analogy with the expression (1) for the molecular viscosity, it is supposed that $\nu_T \sim v_T \Lambda$, where v_T is a characteristic velocity of the turbulent eddies and Λ is the so–called mixing length, which is the "mean free path" of the eddies, i.e. the distance they travel through before they mix with their environment.

The same analogy is used in accretion disk theory through the α model that we shall describe below.

It is important to note that the basis for this analogy is very weak. For a thorough discussion of the differences between molecular and turbulent transport, we refer the reader to Tennekes & Lumley (1972).

In particular, while representing gross transport by an effective viscosity can often be useful, doing a detailed stability analysis on a viscous fluid model for turbulent flow is generally not self–consistent, and can be very misleading (e.g. [13]).

Note that the above discussion applies to *shear* turbulence only, i.e. flows where the fluctuations get their energy from the mean velocity gradients. Transport of momentum may not be nearly as efficient when the energy source for the fluctuations is a gradient of temperature or magnetic field for instance. As a matter of fact, there are strong indications that the transport of momentum associated with thermal convection is orders of magnitude weaker than that associated with shear turbulence ([3] and references therein).

Although turbulence has been considered as a way of transporting angular momentum in accretion disks for more than fifty years, it is only relatively recently that an instability able to extract the energy of the shear and put it in the fluctuations has been elucidated. This instability, which requires a magnetic field, is described in the next section.

3 The Magnetorotational Instability

Three types of waves can propagate in a magnetized fluid: the Alfvén waves, the fast MHD (or magneto–acoustic) waves and the slow MHD waves. The Alfvén waves are transverse (with no motion in the plane defined by the wavenumber and the magnetic field) and propagate along the field lines. They do not involve any compression across the field lines and their restoring force is the magnetic tension. These waves are analogous to the waves which propagate along a stretched string. The fast and slow MHD waves are associated with motions only in the plane defined by the wavenumber and the magnetic field. For the fast mode, the magnetic and thermal pressures act in phase. If the Alfvén speed is large compared to the sound speed, the slow mode is an acoustic wave propagating along the field lines, whereas if the Alfvén speed is small compared to the sound speed, it degenerates into an Alfvén mode in its dispersion properties (the eigenvector is distinct from that of the Alfvén mode however, as the motions are not in the same plane). For the fast mode it is the opposite.

In the absence of rotation, these modes are stable. However, Balbus & Hawley (1991) have shown that when rotation is introduced, the slow mode can become unstable, and this what we describe now.

3.1 Linear Instability

We consider the simplest system in which the instability can develop. This is an axisymmetric disk with a vertical uniform magnetic field. For the original presentation, which includes the case of a radial field, see [1], and for the stability of a toroidal field see [2, 7, 18, 29].

Since it is the slow mode which is destabilized, one can consider an incompressible fluid (in which the fast mode has a frequency which is infinite). The system of equations describing the fluid is then:

$$\frac{\partial \mathbf{B}}{\partial t} = \boldsymbol{\nabla} \times (\mathbf{v} \times \mathbf{B}) \ , \tag{10}$$

$$\boldsymbol{\nabla} \cdot \mathbf{v} = 0 \ , \tag{11}$$

$$\frac{\partial \mathbf{v}}{\partial t} + (\mathbf{v} \cdot \boldsymbol{\nabla}) \mathbf{v} = -\frac{1}{\rho} \boldsymbol{\nabla} P + \frac{1}{\mu_0 \rho} (\boldsymbol{\nabla} \times \mathbf{B}) \times \mathbf{B} - \boldsymbol{\nabla} \Psi \ , \tag{12}$$

where $\mathbf{B}$ is the magnetic field, $\mathbf{v}$ is the fluid velocity, ρ is the density of mass, P is the pressure, Ψ is the gravitational potential due to the central star and μ_0 is the permeability of vacuum.

We use the cylindrical coordinates (r, ϕ, z) and we denote the unit vectors in the three directions by $\mathbf{e}_r$, $\mathbf{e}_\phi$ and $\mathbf{e}_z$. We consider equilibrium quantities that are uniform and a vertical magnetic field. The velocity is $\mathbf{v}_0 = r\Omega(r)\mathbf{e}_\phi$.

We now suppose that this equilibrium is slightly perturbed and we look for solutions proportional to $\exp[i\,(\omega t - k_z z - k_r r)]$. Here we consider axisymmetric perturbations, but more general solutions can be obtained (see the above mentioned papers). We also consider large vertical wavenumbers, such that $|k_z| \gg |k_r|$ and $|k_r| \gg 1/r$. Then, the linearized system of equations leads to the following dispersion relation:

$$\omega^4 - \omega^2 \left(2k^2 v_A^2 + \frac{d\Omega^2}{d\ln r} + 4\Omega^2 \right) + k^2 v_A^2 \left(k^2 v_A^2 + \frac{d\Omega^2}{d\ln r} \right) = 0 \,, \qquad (13)$$

where v_A is the Alfvén speed. Equation (13) is a quartic for ω^2 which solutions are real. Therefore there is instability if $\omega^2 < 0$, which requires:

$$k^2 v_A^2 < -\frac{d\Omega^2}{d\ln r} \,. \qquad (14)$$

This criterion has a very simple physical explanation. It states that there is an instability when the magnetic tension that acts on a segment of a field line is smaller than the net tidal force (i.e. centrifugal force minus gravitational force) acting on it.

For a given equilibrium field B, and therefore a given Alfvén speed v_A, there will always be a wavenumber k such that this inequality is satisfied provided the right hand side is positive. Therefore, all the disks with

$$\frac{d\Omega^2}{d\ln r} < 0 \qquad (15)$$

are unstable, and this is the criterion for instability.

The heart of the instability resides in the fact that a perturbed fluid element tends to conserve its angular velocity when a magnetic field is present. This is to be contrasted with a non magnetized disk, in which a perturbed element tends to conserve its specific angular momentum. When displaced inward therefore, it has too much angular momentum for its new location (as the angular momentum increases outward in an accretion disk), and it moves back to its initial unperturbed position. When a magnetic field is present, the magnetic tension along the field line tends to enforce isorotation of the elements to which it is connected. A fluid element displaced inward has therefore a lower angular velocity than the elements at its new location, and thus not enough angular momentum for its new position. As a result it sinks further in. On the opposite, a fluid element connected to the same field line and displaced outward will tend to move further out. Angular momentum is transferred *via* magnetic torques from the inner fluid element to the outer fluid element. Note that the source of free energy for the instability is not in the magnetic field, but in the disk differential rotation. The magnetic field just provides a path to extract the energy.

From (13), we can write the negative values of ω^2 as a function of $k^2 v_A^2$, and show that the maximum growth rate is:

$$|\omega_{\text{max}}| = \frac{1}{2}\left|\frac{d\Omega}{d\ln r}\right|.$$ (16)

For a Keplerian disk,

$$|\omega_{\text{max}}| = \frac{3}{4}\Omega$$ (17)

and is attained for

$$kv_A = \frac{\sqrt{15}}{4}\Omega.$$ (18)

This holds even if the field has radial and azimuthal components and if the perturbed quantities are allowed to vary with r and ϕ provided we then replace kv_A by $\mathbf{k} \cdot \mathbf{v}_A$. Note that the non–axisymmetric case is more subtle though, as in that case plane waves cannot be sustained, being sheared out by the differential rotation. If we write $\mathbf{k} = k_r\mathbf{e}_r + (m/r)\mathbf{e}_\phi + k_z\mathbf{e}_z$, with m being the azimuthal wavenumber, then k_r is time dependent and

$$k_r(t) = k_0 - mt\frac{d\Omega}{dr}$$

[11], which means that a disturbance always becomes trailing in a disk where the angular velocity decreases outward. If k_r is initially large and positive (leading disturbance), then the mode is stable as indicated by (14). But as time goes on, k_r decreases, so that $\mathbf{k}$ enters a region of instability. As k_r becomes negative and keeps decreasing though, the mode becomes stable again. Formally, the instability is therefore not purely exponential. The important question however is whether the mode can be amplified significantly before its wavelength becomes small enough to be affected by ohmic resistivity. This, of course, depends on the magnetic Reynolds number.

We have seen above that the so–called *magnetorotational* or *Balbus–Hawley instability* can develop in any disk in which the angular velocity decreases outward. In principle there is no other condition. However, it may be that the scale of the modes which are unstable according to (14) do not fit into the disk, i.e. they have a wavelength larger than the disk semithickness H. This is the case if $v_A/\Omega > H$. Since in a thin disk $\Omega H \sim c_s$, the disk will be stable if $v_A > c_s$.

Another condition which is implicit in the above presentation is, of course, that the magnetic field be coupled to the fluid. This may not be the case everywhere in protostellar disks, which are rather cold and dense [9, 10].

3.2 Nonlinear Evolution

Numerical simulations have shown that this instability puts the energy it extracts from the disk differential rotation into fluctuations which transport angular momentum outward (see [3] and references therein). Numerical simulations also show that most of the transport is done by the (magnetic) Maxwell stress, which dominates over the (hydrodynamic) Reynolds stress. Furthermore, magnetic fields are regenerated through a *dynamo* action so that the mechanism can sustain itself in an isolated system.

Long before a mechanism for producing turbulence in accretion disks had been identified, Shakura & Sunyaev (1973) proposed a prescription for modeling turbulent disks. We now describe this prescription, and discuss its validity in the context of magnetic turbulence.

4 Disk Models and Simulations

4.1 Evolution of Turbulent Disks

Here we consider an axisymmetric disk rotating around a central object. We suppose the motion is in the plane of the disk, or, equivalently, we use the vertically averaged equations of mass conservation and motion. The velocity is $\mathbf{v} \equiv (u_r, r\Omega + u_\phi)$. The term $r\Omega$ is the circular velocity around the central mass, and (u_r, u_ϕ) are the components of the fluctuation velocity. Note that as the disk accretes, there is a net radial drift and the mean value of u_r is non zero. The equation of mass conservation and the azimuthal component of the equation of motion are:

$$\frac{\partial \Sigma}{\partial t} + \frac{1}{r}\frac{\partial}{\partial r}\left(r\Sigma v_r\right) = 0 \,, \tag{19}$$

$$\Sigma \left(\frac{\partial v_\phi}{\partial t} + v_r\frac{\partial v_\phi}{\partial r} + \frac{v_r v_\phi}{r}\right) = 0 \,, \tag{20}$$

where Σ is the surface density. Note that we have not included the viscous force arising from molecular viscosity as it is negligible. Multiplying (20) by r and using (19), we obtain the angular momentum equation:

$$\frac{\partial}{\partial t}\left(r\Sigma\left(r\Omega + u_\phi\right)\right) + \frac{1}{r}\frac{\partial}{\partial r}\left[r^2\Sigma\left(r\Omega + u_\phi\right)u_r\right] = 0 \,. \tag{21}$$

As pointed out by Balbus & Papaloizou (1999), to get a diffusion equation describing the disk evolution we need to smooth out the fluctuations over radius. To do so, we average the above equation over a scale large compared to that of the fluctuations, but small compared to the characteristic scale of the flow (i.e. r). Equation (21) then yields:

$$\frac{\partial}{\partial t}\left(\Sigma r^2 \Omega\right) + \frac{1}{r}\frac{\partial}{\partial r}\left(\Sigma r^3 \Omega < u_r > + \Sigma r^2 < u_r u_\phi >\right) = 0 \,, \tag{22}$$

where the brackets denote the radial average and we have neglected $< u_\phi >$ compared to $r\Omega$ in the time derivative. This is justified because $| < u_\phi > | \ll$

$r\Omega$ and the systematic, evolutionary time derivative of $< u_\phi >$ is limited. In the radial divergence term however, both $< u_r >$ and $< u_r u_\phi >$ are second order, and therefore we retain all the terms. This equation is the same as that describing a viscous flow with $\mathbf{v} \equiv (u_r, r\Omega)$ and a stress tensor $\sigma_{r\phi} \equiv -\Sigma < u_r u_\phi >$.

There are two contributions to the flux of angular momentum: the term $\Sigma r^2 \Omega < u_r >$ is the mean angular momentum advected through the disk by the velocity fluctuations (because of the accretion of mass), whereas the term $\Sigma r < u_r u_\phi >$ represents the angular momentum fluctuations transported by the velocity fluctuations.

Using (19) averaged over radius, we can rewrite (22) under the form:

$$r\Sigma < u_r >= -\left[\frac{d}{dr}\left(r^2\Omega\right)\right]^{-1}\frac{\partial}{\partial r}\left(\Sigma r^2 < u_r u_\phi >\right).$$

Using (19) again to eliminate $< u_r >$, this leads to the diffusion equation:

$$\frac{\partial \Sigma}{\partial t} = \frac{1}{r}\frac{\partial}{\partial r}\left\{\left[\frac{d}{dr}\left(r^2\Omega\right)\right]^{-1}\frac{\partial}{\partial r}\left(\Sigma r^2 < u_r u_\phi >\right)\right\}. \qquad (23)$$

In a steady state, (19) gives $r\Sigma < u_r >=$ constant. Then the accretion rate

$$\dot{M} \equiv -2\pi r\Sigma < u_r > \qquad (24)$$

is constant through the disk. Integration of the angular momentum equation (22) then yields:

$$\Sigma < u_r u_\phi >= \frac{\dot{M}}{2\pi}\Omega\left[1 - \left(\frac{R_i}{r}\right)^{1/2}\right], \qquad (25)$$

where R_i is the disk inner boundary. We have assumed here that the turbulent stress $< u_r u_\phi >$ vanishes at the disk inner edge (i.e. there is no torque at the boundary) and that the disk is Keplerian, so that $\Omega \propto r^{-3/2}$. The above equation shows that for the mass to be accreted inward (i.e. $< u_r >$ negative), the flux of angular momentum due to the fluctuations has to be positive, i.e. the fluctuations have to transport angular momentum outward.

4.2 The α Prescription

We pointed out that the angular momentum equation (22) is analogous to that describing a viscous flow with $\mathbf{v} \equiv (u_r, r\Omega)$ and a stress tensor $\sigma_{r\phi} \equiv -\Sigma < u_r u_\phi >$. Therefore, it is tempting to push the analogy further and express the turbulent stress $-\Sigma < u_r u_\phi >$ as if it derived from an enhanced 'turbulent viscosity' ν, defined such that:

$$-\Sigma < u_r u_\phi > \equiv \Sigma \nu r \frac{d\Omega}{dr} \,. \tag{26}$$

In a Keplerian disk, this gives:

$$< u_r u_\phi > \sim \nu \Omega \,. \tag{27}$$

The equations presented in the previous section have been derived by Lynden–Bell & Pringle (1974; see also [22]) assuming a viscous flow with this expression of the stress tensor.

The prescription proposed by Shakura & Sunyaev (1973) consists in writing $\nu = v_t H$, where H is the disk thickness, assumed to be the maximum scale of the turbulent cells, and v_t is the turbulent velocity. They further define

$$\alpha \equiv \frac{v_t}{c_s} \,,$$

where c_s is the sound velocity. Note that $\alpha < 1$, otherwise the fluctuations would dissipate into shocks in such a way as to restore $v_t < c_s$. Equation (27) can then be rewritten under the form:

$$< u_r u_\phi > \sim \alpha c_s^2 \,. \tag{28}$$

Here we have used the fact that in a thin disk $H \sim c_s/\Omega$.

So far we have focussed on non magnetized disks. In these, there are strong indications that $< u_r u_\phi >$ is either zero or negative. Magnetism is needed to correlate the velocities. The above discussion does apply to magnetized disks provided we replace $< u_r u_\phi >$ by $< u_r u_\phi - u_{Ar} u_{A\phi} >$, where $(u_{Ar}, u_{A\phi})$ are the components of the fluctuations of the Alfvén velocity [3, 24]. The extra term represents the Maxwell stress. We then have:

$$\alpha \sim \frac{< u_r u_\phi - u_{Ar} u_{A\phi} >}{c_s^2} \,. \tag{29}$$

4.3 Validity of the α Prescription

The validity of the α prescription in the context of magnetic turbulence was discussed by Balbus & Papaloizou (1999). They first pointed out that, as long as $< u_r u_\phi >$ (or $< u_r u_\phi - u_{Ar} u_{A\phi} >$) is positive, the disk dynamics is the same as if it were evolving under the action of a viscosity. In that case indeed, the diffusion coefficient in (23) is positive. We can then always define an α parameter according to (28), although α may not be constant through the disk.

Note however that this implicitly assumes it is possible to average the equations over radius in the way described in § 4.1. If the scale of the fluctuations and the characteristic disk scale are not well separated, such an average cannot be performed. Since the maximum scale of the fluctuations is of order the disk thickness H, this procedure requires that there is a scale large compared to H and small compared to r. This condition may be only marginally fulfilled in protostellar disks, in which H may be up to 0.1–0.2r.

Balbus & Papaloizou (1999) further noted that for the α prescription to apply, the disk had to behave viscously not only in its *dynamics* but also in its *energetics*. The key point here is that a viscous disk dissipates *locally* the energy it extracts from the shear, whether in a steady state or not. This may not be the case in a turbulent disk where, if the turbulent cascade is not efficient, part of the energy may be advected with the flow. As we have not addressed the energetics of viscous disks above, we will not go into the details of the discussion here. These can be found in Balbus & Papaloizou (1999), who showed that in disks subject to MHD turbulence the energy extracted from the shear is indeed dissipated locally (through the turbulent cascade) whether the disk is evolving or not. Note that this is in general not the case when the turbulence is due to self–gravitating instabilities. In that case indeed, part of the energy is transported by waves through the disk.

4.4 Numerical Simulations

The α parameter in numerical simulations can be calculated using (29). In global simulations of unstratified disks (e.g. [12, 26]), it is found that α is a few times 10^{-3} when the initial imposed magnetic field has zero net flux. This is independent of the geometry of the field. If there is a nonzero net toroidal or vertical magnetic flux, α is a few times 10^{-2} or 10^{-1}, respectively.

It is important to note that in all simulations, the value of α averaged over azimuth and the disk thickness varies locally in time and radius on short timescales, comparable to the rotational timescale. The variations typically reach an order of magnitude [26]. This indicates that the α–disk model is not a good approximation for studying processes that affect the disk on dynamical timescales.

References

1. S. A. Balbus, J. F. Hawley: ApJ, **376**, 214 (1991)
2. S. A. Balbus, J. F. Hawley: ApJ, **400**, 610 (1992)
3. S. A. Balbus, J. F. Hawley: Rev. Mod. Phys., **70**, 1 (1998)
4. S. A. Balbus, J. C. B. Papaloizou: ApJ, **521**, 650 (1999)
5. R. D. Blandford, D. G. Payne: MNRAS, **199**, 883 (1982)
6. C. P. Dullemond, D. Hollenbach, I. Kamp, P. D'Alessio: Models of the structure and evolution of protoplanetary disks. In *Protostars and Planets V*, ed by B. R. Reipurth, D. Jewitt, K. Keil (University of Arizona Press, Tucson, 2006)
7. T. Foglizzo, M. Tagger: A&A, **301**, 293 (1995)
8. J. Frank, A. King, D. Raine: *Accretion Power in Astrophysics* (Cambridge University Press 2002 3rd Edition)
9. S. Fromang, C. Terquem, S. A. Balbus: MNRAS, **329**, 18 (2002)

10. C. F. Gammie: ApJ, **457**, 355 (1996)
11. P. Goldreich, D. Lynden–Bell: MNRAS, **130**, 125 (1965)
12. J. F. Hawley: ApJ, **554**, 534 (2001)
13. J. F. Hawley, S. A. Balbus, J. M. Stone: ApJ, **554**, 49 (2001)
14. M. H. M. Heemskerk, J. C. B. Papaloizou, G. J. Savonije: A&A, **260**, 161 (1992)
15. G. Laughlin, P. Bodenheimer: ApJ, **436**, 335 (1994)
16. D. Lynden–Bell, A. J. Kalnajs: MNRAS, **157**, 1 (1972)
17. D. Lynden–Bell, J. E. Pringle: MNRAS, **168**, 60 (1974)
18. G. I. Ogilvie, J. E. Pringle: MNRAS, **279**, 152 (1996)
19. J. C. B. Papaloizou, D. N. C. Lin: ARA&A, **33**, 505 (1995)
20. J. C. B. Papaloizou, G. J. Savonije: MNRAS, **248**, 35 (1991)
21. B. K. Pickett, P. Cassen, R. H. Durisen, R. Link: ApJ, **504**, 468 (1998)
22. J. E. Pringle: ARA&A, **19**, 137 (1981)
23. B. Reipurth, J. Baly, D. Devine: AJ, **114**, 2708 (1997)
24. N. I. Shakura, R. A. Sunyaev: A&A, **24**, 337 (1973)
25. K. R. Stapelfeldt, J. E. Krist, F. Menard, J. Bouvier, D. L. Padgett, C. J. Burrows: ApJ, **502**, L65 (1998)
26. A. Steinacker, J. C. B. Papaloizou: ApJ, **571**, 413 (2002)
27. H. Tennekes, J. L. Lumley: *A first course in turbulence* (The MIT press 1972)
28. C. E. J. M. L. J. Terquem: Tidally–induced angular momentum transport in disks. In *The Formation of Binary Stars*, Proceedings of IAU Symposium 200, ed by R.D. Mathieu, H. Zinnecker (ASP Conference Series 2001)
29. C. Terquem, J. C. B. Papaloizou: MNRAS, **279**, 767 (1996)
30. C. F. von Weizsäcker: Z. Naturforsch, **3a**, 524 (1948)

Theory of MHD Jets and Outflows

Kanaris Tsinganos

Department of Physics, University of Athens and IASA,
Panepistimiopolis, 157 84 Zografos, Greece
tsingan@phys.uoa.gr

Abstract. A brief review is given of selected results of our analytical and numerical work on the construction of time-independent and time-dependent MHD models for non relativistic astrophysical outflows and jets. The equations for steady MHD plasma flows are first outlined. Next, 1-D spherically symmetric outflows are briefly discussed, namely the Parker thermally driven nonrotating wind, as the classical prototype of all astrophysical outflows and the Weber-Davis magnetocentrifugally driven wind together with its astrophysical implications for magnetic braking, etc. Then, we turn to the 2-D MHD problem for steady and non steady 2-D axisymmetric magnetized and rotating plasma outflows. The only available exact solutions for such outflows are those in separable coordinates, i.e. those with the symmetry of radial or meridional self-similarity. Physically accepted solutions pass from the fast magnetosonic separatrix surface in order to satisfy MHD

K. Tsinganos: *Theory of MHD Jets and Outflows*, Lect. Notes Phys. **723**, 117–159 (2007)
DOI 10.1007/978-3-540-68035-2_6   © Springer-Verlag Berlin Heidelberg 2007

causality. An energetic criterion is outlined for selecting radially expanding winds from cylindrically expanding jets. The basics of jet acceleration, collimation, minimum fieldline inclination and angular momentum removal are illustrated in the context of radially self similar models. Numerical simulations of magnetic self-collimation verify several results of analytical steady solutions. The outflow from solar-type inefficient magnetic rotators is very weakly collimated while that from a ten times faster rotating YSO produces a tightly collimated jet. We also propose a two-component model consisting of a wind outflow from the central object and a faster rotating outflow launched from the surrounding accretion disk which plays the role of the flow collimator. We also briefly discuss the problem of shock formation during the magnetic collimation of wind-type outflows into jets.

Key words: MHD – solar wind – ISM / Stars: jets and outflows – galaxies: jets

1 Introduction

Plasma outflows from the environment of stellar or galactic objects, in the form of uncollimated winds or collimated jets is a widespread phenomenon in astrophysics, Fig. 1. The most dramatic illustration of such highly collimated outflows may be perhaps found in the relatively nearby regions of star

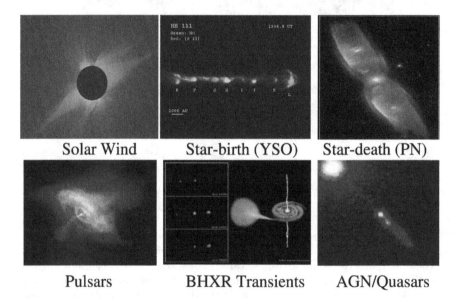

Fig. 1. Large scale MHD astrophysical outflows range from the uncollimated solar wind, to the collimated jets associated with YSO's, planetary nebulae nuclei, pulsars, low- and high-mass X-ray binaries, AGN and quasars. http://sparky.rice.edu/ hartigan/, http://hubblesite.org/, http://chandra.harvard.edu/

formation, Fig. 2; for example, in the Orion Nebula alone the Hubble Space Telescope (HST) has observed hundreds of aligned Herbig-Haro objects [45]. In particular, HST observations show that several jets from young stars are highly collimated within about 30–50 AU from the source star with jet widths of the order of tens of AU, although their initial opening angle is rather large, e.g. $> 60°$, [1, 51]. There is also a long catalogue of jets associated with AGN and possibly supermassive black holes [3]. To a less extend, jets are also associated with older mass losing stars and planetary nebulae [39], symbiotic stars [33], black hole X-ray transients [43], supersoft X-ray sources [34], low- and high-mass X-ray binaries and cataclysmic variables [60]. Even for the two spectacular rings seen with the HST in SN87A, it has been proposed that they may be inscribed by two precessing jets from an object similar to SS433 on a hourglass-shaped cavity which has been created by nonuniform winds of the progenitor star [13, 14]. Also, in the well known long jet of the distant

Fig. 2. Herbig-Haro 49/50, located in the Chamaeleon I star-forming complex, a region containing more than 100 young stars, is believed to be shaped by a cosmic jet packing a powerful punch as it plows through clouds of interstellar gas and dust. The spiral appearance of the jet might be indicative of helical magnetic fields confining the jet. (http://sscws1.ipac.caltech.edu/Imagegallery)

radio galaxy NGC 6251 [32], an about 10^3 light-year-wide warped dust disk perpendicular to the main jet's axis has been observed by HST to surround and reflect UV light from the bright core of the galaxy which probably hosts a black hole [20]. Last but not least, the most powerful sources in our Universe, gamma ray bursts, seem to be associated to the jet phenomenon and are understood as gamma ray synchrotron emission from ultrarelativistic jets.

In the theoretical front, the morphologies of collimated outflows have been studied, to a first approximation, in the framework of ideal steady or time-dependent magnetohydrodynamics (MHD). In *steady* studies, after the pioneering 1-D (spherically symmetric) works of Parker [50], Weber & Davis [79] and Michel [42], Suess [63] and Nery & Suess [44] first modelled the 2-D (axisymmetric) interaction of magnetic fields with rotation in stellar winds, by a linearisation of the MHD equations in inverse Rossby numbers. Although their perturbation expansion is not uniformly convergent but diverges at infinity, they found a poleward deflection of the streamlines of the solar wind caused by the toroidal magnetic field. Blandford & Payne [4] subsequently demonstrated that astrophysical jets may be accelerated magnetocentrifugally from Keplerian accretion disks, *if* the polodial fieldlines are inclined by an angle of 60°, or less, to the disk midplane (but see also, (17)). This study introduced the often used "bead on a rotating rigid wire" picture, although these solutions are limited by the fact that they contain singularities along the system's axis and also terminate at finite heights above the disk, [26, 75]. Sakurai [53] extended the Weber & Davis [79] equatorial solution to all space around the star by iterating numerically between the Bernoulli and transfield equations; thus, a polewards deflection of the poloidal fieldlines was found not only in an initially radial magnetic field geometry, but also in a split-monopole one appropriate to disk-winds, [54]. The methodology of meridionally self-similar exact MHD solutions with a variable polytropic index was first introduced by Low & Tsinganos [40, 66] in an effort to model the *heated* axisymmetric solar wind. Heyvaerts & Norman [30, 31] have shown analytically that the asymptotics of a particular fieldline in non isothermal polytropic outflows is parabolic if it does not enclose a net current to infinity; and, if a fieldline exists which does enclose a net current to infinity, then, somewhere in the flow there exists a cylindrically collimated core. Later, Bogovalov [6] showed analytically that there *always* exists a fieldline in the outflowing part of a *rotating* magnetosphere which encloses a finite total poloidal current and therefore the asymptotics of the outflow *always* contains a cylindrically collimated core. In that connection, it has been shown in [5] that the poloidal fieldlines are deflected towards the polar axis for the split monopole geometry and relativistic or nonrelativistic speeds of the outflowing plasma. Sauty & Tsinganos [55] have self-consistently determined the shape of the fieldlines from the base of the outflow to infinity for nonpolytropic cases and provided a simple *criterion* for the transition of their asymptotical shape from conical (in *inefficient* magnetic rotators) to cylindrical (in *efficient* magnetic rotators). They have also conjectured that as a young star spins down loosing angular momentum, its collimated jet-type outflow becomes gradually a conically expanding wind. Nevertheless, the degree of the collimation of the solar wind at *large* heliocentric distances remains still observationally unconfirmed, since spacecraft observations still offer ambiguous evidence on this question.

Another interesting property of collimated outflows has emerged from studies of various self-similar solutions, namely that in a large portion of them cylindrical collimation is obtained only after some oscillations of decaying amplitude in the jet-width appear [73]. In a series of papers, Sauty and collaborators [56, 57, 58, 59] have elaborated the physics of acceleration and collimation of MHD outflows by studying quasi-analytically *meridionally* self-similar outflows. Similarly, Vlahakis et al. [75] have presented the first quasi-analytical example of a *radially* self-similar outflow which crosses the fast magnetosonic separatrix. The general properties of these MHD separatrices of the MHD equations have been discussed by Tsinganos and collaborators [68]. Radially self-similar models with cylindrical asymptotics for self-collimated and magnetically dominated outflows from accretion disks have been also constructed [46]. All existing cases of self-similar, jet- or, wind-type exact MHD solutions have been unified Vlahakis & Tsinganos [74] in a systematic analytical treatment wherein all available today examples of exact solutions emerge as special cases of a general formulation while at the same time new families with various asymptotical shapes, with (or without) oscillatory behaviour emerge as a byproduct of this systematic method. Altogether, some general trends on the behaviour of stationary, analytic, axisymmetric MHD solutions for MHD outflows seem to be well at hand.

However, observations seem to indicate that jets may inherently be variable. Thus, time-dependent simulations may be useful for a detailed comparison with the observations. Uchida & Shibata [71] were the first to perform time-dependent simulations and demonstrate that a vertical disk magnetic field if twisted by the rotation of the disk can launch bipolar plasma ejections through the torsional Alfvén waves it generates. However, this mechanism applies to fully episodic plasma ejections and no final stationary state is reached to be compared with stationary studies. Similar numerical simulations of episodic outflows from Keplerian disks driven by torsional Alfvén waves on an initially vertical magnetic field have been presented in Goodson et al. [47, 48]. [27] have proposed a time-dependent jet launching and collimating mechanism which produces a two-component outflow: hot, well collimated jet around the rotation axis and a cool but slower disk-wind. Numerical viscosity however results in nonparallel flow and magnetic fields in the poloidal plane in the limited grid space of integration. Washimi & Shibata [78] modelled axisymmetric thermo-centrifugal winds with a *dipole* magnetic flux distribution $B_p^2(\theta) \propto (3\cos^2\theta + 1)$ on the stellar surface (and a radial field in [77]). In this case the magnetic pressure distribution varies approximately as $B_\phi^2 \propto B_p^2 \sin^2\theta$ such that it has a maximum at about $cos^{-1}2\theta_o \approx -1/3$, or, $\theta_o \approx 55°$. As a result, the flow and flux is directed towards the pole and the equator from the midlatitudes around θ_o. The study was performed for uniform in latitude rotation rates and up to 60 solar radii in the equatorial plane. Bogovalov [7, 8] modelled numerically the effects of the Lorentz force in accelerating and collimating a cold plasma with an initially monopole-type magnetic field, in a region limited also by computer time, i.e. the near zone to the central spherical object. In a series of papers, Bogovalov & Tsinganos [9, 10, 11, 69, 70] have used a novel way to model numerically MHD outflows to very large distances by first integrating the time-dependent MHD equations in the near zone containing the critical surfaces and then extending these solutions to unlimited distances along the jet in the superfast regime. In this review we shall constrain our attention only on jets and not to their associated accretion disks. For example, in [38], the reader may find more on a global jet/circulation model for young stars, or, in [19, 37] more on laboratory plasma jets and magnetic tower outflows from a radial wire array in the Z-pinch. We do not intend to discuss instabilities in jets and the interested reader may see, e.g. [35] for pressure and magnetic shear-driven instabilities and [64] for Kelvin-Helmholtz instabilities in rotating MHD jets. We further refer the interested reader to the review article on accretion disks by J. Ferreira in this Volume and references therein, e.g. [22, 23], or other reviews on the subject, e.g. [15, 24, 28, 41, 52].

This review lecture is organised as follows. Section 2 briefly outlines the governing MHD equations and Sect. 3 the basics of 1-D HD outflows, i.e. the classical solar wind theory. Then, Sect. 4 reviews 1-D rotating MHD outflows introducing the concepts of MHD critical points, angular momentum loss via magnetized winds and slow/fast magnetic rotators. Section 5 deals with some simple physical results pertinent to disk winds, such as the acceleration and collimation of the jet, the minimum inclination angle of the fieldlines in the disk midplane and the rate of angular momentum removal from the disk. Section 6 deals with 2-D steady MHD outflows and in particular the various classes of self-similar solutions, the nature of the MHD separatrices selecting a physically acceptable solution and the criteria for collimation. Finally, in Sect. 7 we turn to some recent results on numerical simulations of MHD outflows and give a simple example to explain why magnetized and rotating jets are collimated. At the end we and briefly outline our two-component model which we believe describes appropriately cosmical jets.

2 MHD Equations

The interaction of magnetized plasmas is governed by the familiar MHD equations (Figs. 3, 4). They consist of Maxwell's equations:

$$\boldsymbol{\nabla} \cdot \boldsymbol{E} = 4\pi\delta \approx 0\,, \quad \boldsymbol{\nabla} \cdot \boldsymbol{B} = 0\,, \tag{1}$$

$$\boldsymbol{\nabla} \times \boldsymbol{E} = -\frac{1}{c}\frac{\partial \boldsymbol{B}}{\partial t}\,, \quad \boldsymbol{\nabla} \times \boldsymbol{B} = \frac{4\pi}{c}\boldsymbol{J} + \frac{1}{c}\frac{\partial \boldsymbol{E}}{\partial t} \simeq \frac{4\pi}{c}\boldsymbol{J}\,, \tag{2}$$

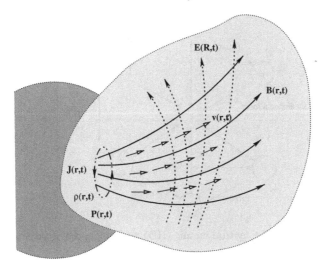

Fig. 3. Illustration of the interaction of magnetized and rotating plasmas with density and pressure (ρ, P) in hydromagnetic fields, $(\boldsymbol{V}, \boldsymbol{B})$

coupled to Ohm's law for a plasma of very high electrical conductivity σ,

$$\boldsymbol{E} + \frac{\boldsymbol{V} \times \boldsymbol{B}}{c} = \frac{\boldsymbol{J}}{\sigma} \simeq 0 \,, \tag{3}$$

Newton's law expressing conservation of angular momentum,

$$\rho \frac{\partial \boldsymbol{V}}{\partial t} + (\rho \boldsymbol{V} \cdot \boldsymbol{\nabla}) \boldsymbol{V} = -\boldsymbol{\nabla} P + \frac{\boldsymbol{J} \times \boldsymbol{B}}{c} - \rho \boldsymbol{G} \,, \tag{4}$$

the continuity equation expressing conservation of mass,

$$\boldsymbol{\nabla} \cdot \rho \boldsymbol{V} + \frac{\partial \rho}{\partial t} = 0 \,, \tag{5}$$

and finally an equation for energy conservation,

$$\rho \left[P \frac{\mathrm{d}}{\mathrm{d}t} \left(\frac{1}{\rho} \right) + \frac{\mathrm{d}}{\mathrm{d}t} \left(\frac{P}{\rho(\Gamma - 1)} \right) \right] = \rho \frac{\partial h}{\partial t} - \frac{\partial P}{\partial t} + \rho \boldsymbol{V} \cdot \left[\nabla h - \frac{\nabla P}{\rho} \right] = q \,. \tag{6}$$

The symbols have their usual meaning, i.e. $\boldsymbol{V}(x_1, x_2, x_3, t)$, $\boldsymbol{B}(x_1, x_2, x_3, t)$ are the bulk flow speed and magnetic field in the plasma which is generated by an electric current with surface density $\boldsymbol{J}(x_1, x_2, x_3, t)$, $\boldsymbol{G}(x_1, x_2, x_3)$ is the external gravitational field, $\rho(x_1, x_2, x_3, t)$ and $P(x_1, x_2, x_3, t)$ are the plasma density and pressure, $h(x_1, x_2, x_3, t)$ the enthalpy of the gas $[= (\Gamma/\Gamma - 1)(P/\rho)]$ and finally $q(x_1, x_2, x_3, t)$ denotes the volumetric rate of energy addition, all expressed in some curvilinear coordinates (x_1, x_2, x_3).

Note that the above set of equations is a rather simplified version of the original full MHD equations for ideal plasmas which describe to a zeroth

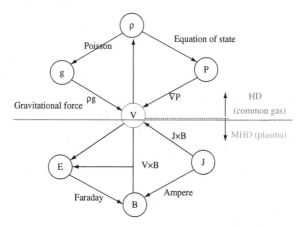

Fig. 4. Illustration of the hydrodynamic (HD) and magnetohydrodynamic (MHD) interaction

order approximation the hydromagnetic interaction in typical astrophysical plasmas, such as those encountered in stellar atmospheres and star formation regions. In such cases, the following approximations hold:

- for nonrelativistic motion, such as is the case in jets associated with star forming regions, the bulk flow speed is much less than the speed of light, $V/c \ll 1$.
- the Spitzer electrical conductivity is rather high for the typical relatively high astrophysical temperatures of thousands to millions of degrees, $\sigma \simeq 6 \times 10^6 T^{3/2}/sec$ wherein the plasma rather behaves like a superconductor.
- from Ohm's law in the limit of high electrical conductivity it follows that the electric field in an inertial frame is,

$$J = \sigma\left[E + \frac{V}{c} \times B\right], \ \sigma \to \infty \ \Rightarrow E + \frac{V}{c} \times B \simeq 0 \Longrightarrow E = -\frac{V}{c} \times B$$

with the result that the displacement current in Ampere's law is negligible. To see that, denote by $(\ell - \tau)$ the characteristic spatial-temporal scales in the system such that $\tau \simeq \ell/V$. Then, since the electric field is of order $E \sim (V/c)B$ it follows that Maxwell's displacement current is

$$\frac{1}{c}\frac{\partial E}{\partial t} \simeq \frac{VB}{c\tau} = \frac{V^2}{c}\frac{B}{\ell} \simeq \frac{cB}{\ell}\left(\frac{V}{c}\right)^2,$$

i.e. of the small order V^2/c^2 with respect to the conduction current which is of the order $c|\nabla \times B| \simeq cB/\ell$.

- the polarization current $J_\delta = \delta V$ associated with deviations from neutrality in $\nabla \cdot E = 4\pi\delta$ is negligible, $J_\delta = V\delta = VE/\ell = (cB/\ell)(V^2/c^2)$ being also of order V^2/c^2 compared to the conduction current cB/ℓ.

- the thermal conductivity κ of the gas is also high for the typical relatively high astrophysical temperatures of thousands to millions of degrees, $\kappa \simeq 6 \times 10^{-6}T^{5/2}erg/cm°K sec$. This has the result, for example, that the solar corona extends far in the interplanetary space, as we shall see in the next section.
- the viscosity (mainly from the ions) is also negligible for the diluted astrophysical gases, $\mu \simeq 10^{-16}T^{5/2}\,g/cm\,sec$, a result highlighting the need for an anomalous viscosity in order to account for angular momentum transfer in astrophysical discs.

As an illustrative example, consider a plasma with temperature $T \sim 10^6 K$ and density $n \sim 10^8\,cm^{-3}$, i.e. values encountered in a stellar corona. The electrical conductivity is comparable to that of Copper ($\sigma_{Cu} \simeq 10^{16}/sec$), while the thermal conductivity is also high $\chi = (\kappa/c_p) \simeq 6 \times 10^8/10^9 \sim 0.6\ g/cmsec$ and the viscosity negligible, $\mu \simeq 0.1\,g/cm\,sec$. The Debye length λ_D is of the order of a cm $[\lambda_D = 6.9(T/n)^{1/2}\,cm]$, much smaller than the typical dimension ℓ of such systems and thus quasi-neutrality holds. The Larmor radius of the protons $r_L = 10^{-4}V_\perp/B\,cm$ for a 1 Gauss magnetic field and a few hundred km/sec bulk flow speed V is about 10 m, while the collisional Coulomb mean free path is $\lambda = 3 \times 10^{-12}V^4/n \approx 3000\ km$. Then, since $\ell \gg \lambda \gg r_L \gg \lambda_D$ and also the number of electrons inside a Debye sphere is large (of the order of 10^9), the fluid approximation is very good and MHD provides an excellent description of the plasma.

Similar is the situation for the plasma of YSO jets, although their temperature, density and ionization fraction are lower, e.g. $T \sim (10^3 - 10^4)K$, $n \sim (10^2 - 10^3)\,cm^{-3}$, while observations show that optical HH jets are moderately ionized with ionization fractions $n_i/n_n = (0.5 - 0.01)$ and lower values at the optically invisible parts of the jet [2, 25]. Nevertheless, again the dimension ℓ of the system (with jet radii R_j of the order of several AU), far exceeds the corresponding Debye length $[\lambda_D = 6.9(T/n)^{1/2}\,cm$, i.e. of the order of 1 $m]$ while the number of electrons inside a Debye sphere N_D is still large $[N_D = n4\pi\lambda_D^3/3$, i.e. of the order of $10^7]$. The collisional Coulomb mean free path $[\lambda = 3 \times 10^{-12}V^4/n\,cm]$ may be of the order of an AU, but as in the case of the solar wind, the proton Larmor radii $[r_L = 10^{-4}V_\perp/B\,cm]$ are much smaller and thus the magnetic field ties the charged particles close to each other. Then again $\ell \gg \lambda \gg r_L \gg \lambda_D$ holds and the fluid approximation is a very good one, concluding that MHD can still provide the best zeroth order description of the dynamics of this plasma as well.

3 The Parker Thermally Driven, Nonrotating Wind

It is beyond doubt that Parker's [49] elegant theory of the solar wind plays a unifying and central role in the understanding of large scale astrophysical plasma outflows. In addition, this theory remains today as one of the most

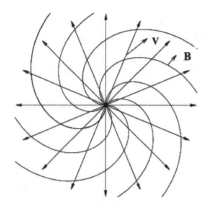 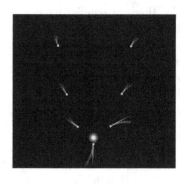

Fig. 5. Solar winds's hydromagnetic field (*left*) and the antisolar alignment of comet tails (*right*)

beautiful topics of theoretical astrophysics being at the same time a classical theory of striking mathematical simplicity which nevertheless is able to explain a rather complex and wide spread astrophysical phenomenon.

3.1 Basic Facts Leading to the Solar Wind

The basic facts that underline the existence of the Parker solar wind are:

- The high temperature T of the solar corona ($T \approx 10^6 - 10^{7\circ}K$) combined with the low temperature of the interstellar medium (ISM) (T $\approx 100^\circ K$).

- The high thermal conductivity κ of the plasma: $\boldsymbol{q} = -\kappa(T)\boldsymbol{\nabla}T$.

  For plasmas of such high T, the thermal conductivity coefficient is given by $\kappa(T) \approx \times 10^{-13}\, T^{5/2}$cal/cm sec K.

  At typical coronal temperatures $T = 10^7 K \longrightarrow \kappa \approx 1$cal/cm sec K

  (note that for Copper, Silver: $\kappa \approx 1$cal/cm sec K)

- Because of this high value of κ the temperature T should be homogenized:

  $$\boldsymbol{\nabla}\cdot[\kappa(T)\boldsymbol{\nabla}T] = 0 \implies T = \frac{T_o}{R^{2/7}}$$

  At $R \approx 10^4\,(ISM) \implies T = \frac{10^6}{10^{8/7}} >> 100\,K.$

  It follows that inevitably the atmosphere expands.

- Formally, this can be also seen from Bernoulli's law:

  $$\frac{kT_o}{\gamma - 1} + \frac{V_o^2}{2} - \frac{GM}{r_o} = \frac{kT_\infty}{\gamma - 1} + \frac{V_\infty^2}{2} - \frac{GM}{r_\infty},$$

taking into account that in the LHS the positive sum of the enthalpy and the gravitational potential is asymptotically transferred to kinetic energy of the plasma in the RHS where the gravitational and enthalpic terms are negligible. In other words, we have eventually a transfer of random thermal energy (enthalpy) from the base of the outflow to directed kinetic energy of the plasma further away.

This summarizes the physics of the solar wind phenomenon which was proposed theoretically by Parker in 1958 [49] and was observed by the spacecraft Mariner 2 in 1962.

3.2 Parker's Isothermal Wind Theory

The basic equations of Parker's simplified solar wind theory are:

$$\frac{d}{dr}\left(\rho V r^2\right) = 0 \text{ (mass conservation)}, \tag{7}$$

$$\rho V \frac{dV}{dr} = -\frac{dP}{dr} - \rho \frac{GM}{r^2} \text{ (momentum conservation)}, \tag{8}$$

$$P = \rho V_s^2, \quad V_s = \sqrt{\frac{2kT_0}{m}} \text{ (equation of state)}, \tag{9}$$

where V_s is the sound speed, i.e. the most probable proton speed in a Maxwellian distribution with temperature T_0. The dimensionless distance R and density ρ are given in units of the base radius r_0 and base density ρ_0,

$$R = \frac{r}{r_0}, \quad M = \frac{V}{V_s}. \tag{10}$$

In terms of the dimensionless parameter λ,

$$\lambda = \frac{1}{2}\left(\frac{V_{esc}}{V_s}\right)^2 = \frac{GMm}{2r_0kT_0} \simeq 12 \text{ for } T_0 \simeq 10^6\,^\circ K, V_s = 131 \text{ km/s}, \tag{11}$$

we may combine the above Equation to obtain two equations for $\rho(R)$ and $M(R)$:

$$\frac{d}{dR}\left(\rho M R^2\right) = 0 \implies \rho M R^2 = \mu = \text{const}.$$

$$\rho M \frac{dM}{dR} + \frac{d\rho}{dR} + \lambda \frac{\rho}{R^2} = 0 \implies \frac{1}{M^2}\frac{dM^2}{dR} = \frac{2}{R^2}\frac{2R - \lambda}{M^2 - 1}. \tag{12}$$

The last equation can be integrated to give Bernoulli's integral:

$$\ln M - \frac{M^2}{2} - \frac{\lambda}{R} + \ln R^2 = \ln B = \text{const}. \tag{13}$$

The momentum equation (12) has a critical point at $2R_c = \lambda$, $M = 1$, Fig. 7, wherein $dM^2/dR = 0/0$. In the vicinity of this critical point we may write $R = R_c(1 + \varepsilon)$, $M = M_c(1 + \xi)$. Figures 6, 7 illustrate the characteristic saddle-type topology around this singularity. Note that exactly the same procedure can be applied to the case of polytropic winds with $\gamma \neq 1$ with similar results, i.e. we obtain an analogous Bernoulli integral, critical point and topology of the solutions in the plane (M,R). In the following, we briefly outline a systematic procedure for obtaining the critical points of the wind problem, since this may be also followed in the case of polytropic magnetized winds discussed in the next section.

From the combination of the two integrals expressing energy and mass conservation, we may obtain the energy as a function of the two variables of the radial distance r and the density ρ:

$$\left.\begin{array}{l} \mathcal{E} = \frac{1}{2}V^2 + \frac{\gamma}{\gamma-1}\rho^{\gamma-1} - \frac{GM}{r}\,, \\[2mm] \rho V r^2 = \mathcal{F}_m = \text{const}., \end{array}\right\} \Rightarrow$$

$$\mathcal{E} = \frac{\mathcal{F}_m^2}{2\rho^2} + \frac{\gamma}{\gamma-1}k\rho^{\gamma-1} - \frac{GM}{r} = \mathcal{E}(\rho,r)\,.$$

Then, since, $\mathcal{E}(\rho,r) = \text{const}. \Rightarrow d\mathcal{E} = 0$ and for a finite density gradient at the critical point we require that:

$$\frac{d\rho}{dr} = \frac{\left.\frac{\partial \mathcal{E}}{\partial r}\right|_\rho}{\left.\frac{\partial \mathcal{E}}{\partial \rho}\right|_r} = \text{finite} \implies \left.\frac{\partial \mathcal{E}}{\partial r}\right|_\rho = \left.\frac{\partial \mathcal{E}}{\partial \rho}\right|_r = 0\,. \tag{14}$$

The first equation determines the location of the critical point and the second the corresponding critical speed:

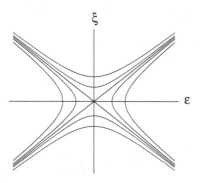

Fig. 6. X-type critical point in the topology of the wind solutions

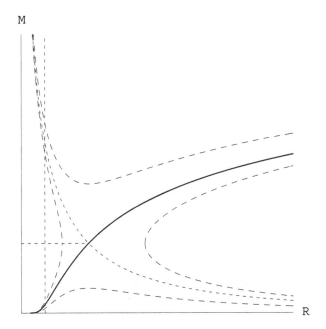

Fig. 7. Topology of the Bernoulli integral. The physically accepted solution is indicated with the thick line

$$
\left\{
\begin{array}{l}
\rho \left. \dfrac{\partial \mathcal{E}}{\partial \rho} \right|_r = -V^2 + \gamma k \rho^{\gamma-1} = -V_{cr}^2 + V_s^2 = 0, \\[3mm]
r \left. \dfrac{\partial \mathcal{E}}{\partial r} \right|_\rho = -2V_{cr}^2 + \dfrac{GM}{r_{cr}} = 0,
\end{array}
\right\}
\Rightarrow
\left\{
\begin{array}{l}
V_{cr} = V_s, \\[3mm]
r_{cr} = \dfrac{GM}{2V_s^2},
\end{array}
\right.
$$

where V_s denotes the sound speed and V_{cr} the speed at the critical point at the distance r_{cr}.

4 Rotating, Magnetized Equatorial Winds

The simplest model of a rotating and magnetized wind which was proposed by Weber & Davis in 1967, [79] nicely explains the magnetic braking of stars. From the formal point of view, this model also reveals the more complicated topology of the solutions of an MHD outflow which now need to cross the various MHD critical points, instead of the single sonic critical point encountered in Parker's theory.

4.1 Basic Equations

In the Weber and Davis (WD) model we solve the MHD equations on the equatorial plane $\theta = \pi/2$ of the rotating star by using spherical coordinates

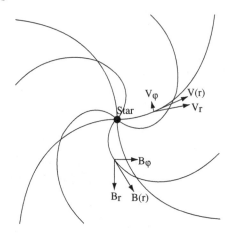

Fig. 8. The Weber-Davis magnetized and rotating wind

(r, ϑ, φ) as depicted in Fig. 8. A drastic simplification results by ignoring the meridional components along $\hat{\theta}$, and also the dependence of the variables on the azimuthal angle φ. Thus, the five assumptions of the WD model of a magnetocentrifugal wind are:

- *Steady* states, $\partial_t = 0$.
- *Axisymmetry* around the z-axis, $\partial_\varphi = 0$.
- Analysis of the problem on the *equatorial* plane, $\vartheta = \pi/2$.
- *Neglect the meridional components* of the hydromagnetic field $(\boldsymbol{V}, \boldsymbol{B})$ which are assumed to only have radial and azimuthal components depending only on the radial distance r,

$$\boldsymbol{V} = \boldsymbol{V}_r(r) + \boldsymbol{V}_\varphi(r) \,,$$

$$\boldsymbol{B} = \boldsymbol{B}_r(r) + \boldsymbol{B}_\varphi(r) \,,$$

- A *polytropic* relation between pressure and density, $P = k\rho^\gamma$.

Also, in the model the following physical quantities are involved:

- the Alfvén speed in the radial direction, $V_{Ar} = \dfrac{B_r(r)}{\sqrt{4\pi\rho(r)}}$,
- the Alfvén number M_r for magnetized fluids, a quantity equivalent to the Mach number for HD winds,

$$M_r = \frac{V_r(r)}{V_{Ar}(r)} \equiv M(r) \,,$$

- the Alfvén radius $r_\star$, i.e. the radial distance in which the radial outflow speed $V_r(r)$ equals to the radial Alfvén speed V_{Ar}, $M(r_\star) = 1$,
- the Alfvén density $\rho_\star$, i.e. the plasma density at the Alfvén distance: $\rho(r_\star) = \rho_\star$,

- the radial speed $V_\star$ at the Alfvén distance, $V_\star = V_{Ar}(r_\star)$,
- the adimensional radial distance in units of the Alfvén radius , $R = \frac{r}{r_\star}$,

- the escape speed at the Alfvén radius, $V_{esc}(r_\star) = \sqrt{\frac{2G\mathcal{M}}{r_\star}}$.

From these quantities one may define the following adimensional constants, by normalising the escape speed V_{esc}, the sound speed V_s and the rotational speed at the Alfvén distance $r_\star$, $\Omega r_\star$, with the Alfvén speed $V_\star$ at the radius $r_\star$, and finally the total energy per unit mass $\mathcal{E}$ in units of $V_\star^2/2$. Thus we have:

- the adimensional escape speed, $\nu^2 = \dfrac{V_{esc}^2}{V_\star^2} = \dfrac{2G\mathcal{M}}{r_\star V_\star^2}$,
- the adimensional sound speed, $s^2 = \dfrac{2\gamma}{\gamma-1}\dfrac{V_s^2(r_\star)}{V_\star^2} = \dfrac{2\gamma k\rho_\star^{\gamma-1}}{(\gamma-1)V_\star^2}$,
- the adimensional rotational speed at the Alfvén radius, $\omega = \dfrac{\Omega r_\star}{V_\star}$,
- the adimensional total energy per unit mass, $\epsilon = \dfrac{2\mathcal{E}}{V_\star^2}$.

4.2 Azimuthal Components of the Hydromagnetic Field

In the following we briefly outline the steps leading to the expressions of the azimuthal magnetic field and the flow. Since the hydromagnetic field is assumed to only have radial and azimuthal components only, the induction electric field is $\boldsymbol{E} = E_\vartheta \boldsymbol{\vartheta}$. Then, from Faraday's induction law

$$\boldsymbol{\nabla} \times \boldsymbol{E} = \frac{\hat{\varphi}}{r}\frac{\partial}{\partial r}(rE_\vartheta) = 0 \, ,$$

we have,

$$-crE_\vartheta = r(\boldsymbol{V} \times \boldsymbol{B}) \cdot \hat{\boldsymbol{\vartheta}} = r(V_\varphi B_r - V_r B_\varphi) \approx \Omega r^2 B_r = \Omega r_0^2 B_r(r_0) \, , \quad (15)$$

since at the base r_0, where $V_\varphi = \Omega r_0$ the radial speed $V_r(r_0)$ is negligible. The azimuthal component of the momentum equation gives:

$$[\rho(\boldsymbol{V} \cdot \boldsymbol{\nabla})\boldsymbol{V}] \cdot \hat{\varphi} = [\frac{(\boldsymbol{\nabla} \times \boldsymbol{B}) \times \boldsymbol{B}}{4\pi}] \cdot \hat{\varphi} \implies \frac{\partial}{\partial r}(rV_\varphi) = \frac{B_r}{4\pi\rho V_r}\frac{\partial}{\partial r}(rB_\varphi) \, (16)$$

But the fluxes $F_B = r^2 B_r$ and $F_m = \rho r^2 V_r$ are constants and so,

$$\frac{4\pi\rho r^2 V_r}{r^2 B_r} = \frac{F_m}{F_B} \equiv \Psi_A = \text{const} \implies rV_\varphi - \frac{rB_\varphi}{\Psi_A} = L \, , \quad (17)$$

where L is the specific total angular momentum of the plasma and the electromagnetic field due to the magnetic torque of the Lorentz force $cF_\varphi = J_\vartheta B_r$ around the z-axis. Combining (16) and (17) we get:

$$V_\phi(r) = \frac{L}{r} \frac{M^2 - \frac{\Omega}{L} r^2}{M^2 - 1}, \quad \frac{B_\phi(r)}{\Psi_A} = \frac{L}{r} \frac{1 - \frac{\Omega}{L} r^2}{M^2 - 1}. \tag{18}$$

At the Alfvenic distance $r = r_*$, $M = 1$, and in order that the expressions are finite we need to have $\Omega r_*^2 = L$. Finally, in terms of $R \equiv r/r_*$ we have:

$$V_\phi(R) = \frac{\Omega r_*}{R} \frac{M^2 - R^2}{M^2 - 1}, \quad \frac{B_\phi(R)}{\Psi_A} = \frac{\Omega r_*}{R} \frac{1 - R^2}{M^2 - 1}. \tag{19}$$

This completes the derivation of the azimuthal components of the hydromagnetic speed in terms of the radial distance R and the Alfvén number $M(R)$ and also the two constants Ω and Ψ_A which determine the angular speed of the roots of the fieldlines on the star or surrounding disk and the fixed ratio of mass to magnetic fluxes.

4.3 The Generalized Bernoulli Integral

By substituting the above expressions of the azimuthal fields in the radial component of the momentum equation we get,

$$\frac{d \ln V_r}{d \ln R} = \frac{(V_r^2 - V^2)\left(2V_s^2 + V_\phi^2 - \frac{GM}{r_* R}\right) + 2V_r V_\phi V_A V_{A\phi}}{(V_r^2 - V_{sl}^2)(V_r^2 - V_f^2)}. \tag{20}$$

Note the two critical points in this equation which correspond to the slow V_{slow} and fast V_{fast} characteristic MHD wave speeds.

Integrating the above equation we obtain similarly to the Parker wind the generalized energy integral (Bernoulli equation);

$$\mathcal{E} = \underbrace{\frac{1}{2}V_r^2 - \frac{GM}{r} + h}_{\substack{\text{hydrodynamic} \\ \text{term (Parker)}}} + \underbrace{\frac{1}{2}V_\phi^2 - \frac{rB_\phi \Omega}{\Psi_A}}_{\substack{\text{new magnetic -} \\ \text{rotational term}}} = \text{const}, \quad h = \frac{\gamma}{\gamma - 1} k \rho^{\gamma - 1}. \tag{21}$$

A plot of the Bernoulli integral is shown in the following Fig. 9 while Fig. 10 shows a typical solution.

The generalized Bernoulli integral can be written as:

$$\mathcal{E} = \frac{1}{2}V^2 + h - \frac{GM}{r} - \frac{rB_\phi \Omega}{\Psi_A} = \underbrace{\frac{1}{2}V_o^2 + h_o - \frac{GM}{r_o}}_{\mathcal{E}_o} - \underbrace{\frac{r_o B_\phi^\circ \Omega}{\Psi_A}}_{\approx \Omega L} \approx \mathcal{E}_o + \Omega L \tag{22}$$

where $\mathcal{E}_o$ is the specific energy of the thermally driven Parker wind and ΩL is the Poynting energy of the magnetic rotator. Depending on which of these two terms dominates we have two possibilities:

1. $\mathcal{E}_o \gg \Omega L$: **Slow magnetic rotator, SMR**. In this case we have a thermally driven Parker wind
2. $\mathcal{E}_o \ll \Omega L$: **Fast magnetic rotator, FMR**. In this case we have a magnetorotationally driven wind.

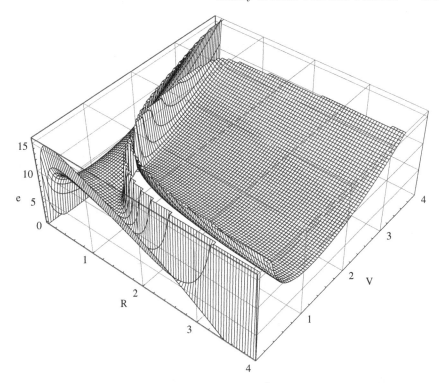

Fig. 9. 3-D plot of the Bernoulli integral $\epsilon = 2\mathcal{E}/V_*^2$ as a function of the dimensionless velocity $u = V_r/V_*$ and radius $R = r/r_*$

4.4 The Michel Characteristic Speed

Since the enthalpic and gravitational terms become negligible at large distances, the Bernoulli equation for a magnetically dominated wind becomes asymptotically,

$$-\left.\frac{\Omega r B_\phi}{\Psi_A}\right|_{r\to\infty} = -\Omega^2 r^2 \frac{1-R^2}{M^2-1} \simeq \Omega^2 r_*^2 \frac{R^2}{M^2} = \frac{\Omega^2 r_*^2}{u_\infty} , \qquad (23)$$

$$\mathcal{E} \approx \frac{1}{2}V_\infty^2 - \left.\frac{\Omega r B_\phi}{\Psi_A}\right|_\infty , \qquad (24)$$

where $u = V_r/V_* = M^2/R^2$. In terms of the dimensionless rotational speed ω at the Alfvenic distance, $\omega = \Omega r_*/V_*$ we have at these large distances,

$$\epsilon(u) = \frac{2\mathcal{E}(u)}{V_*^2} = u^2 + \frac{2\omega^2}{u} . \qquad (25)$$

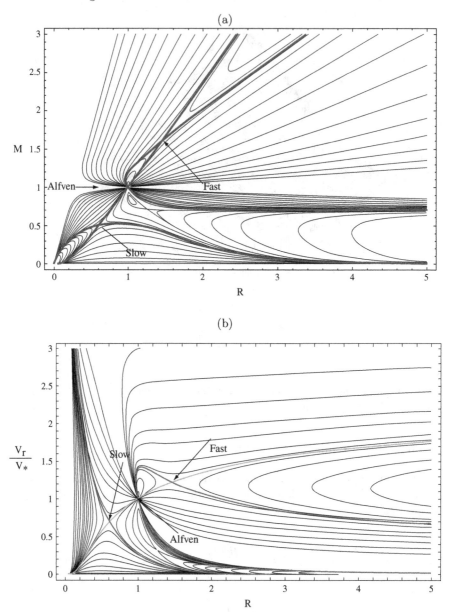

Fig. 10. A typical solution for the parameters $\nu = 1.401256, s = 2.84215, \omega = 1.016858, \gamma = 1.133537$, corresponding to a slow magnetic rotator. With thick solid lines are indicated the solutions which cross the critical points, slow, Alfvén and fast. The solutions are shown in the plane $M - R$ in (a) and the plane $u - R$ in (b), with $u = V_r/V_* = M^2/R^2$

The double root of this equation - the minimum of the curve $\epsilon(u) \equiv f(u)$ see Fig. 12 - is $\epsilon(u_0) = \epsilon_0$ at $u_o = \omega^{2/3}$:

$$u_0 = \frac{V_M}{V_\star} = \omega^{2/3} \Rightarrow V_M^3 = V_\star^3 \left(\frac{\Omega^2 r_\star^2}{V_\star^2} \right) = \frac{\Omega^2 r_\star^2}{V_\star} \frac{B_\star^2}{4\pi\rho_\star} ,$$

$$\mathcal{F}_B = B_{r,\star} r_\star^2 , \qquad \mathcal{F}_m = 4\pi\rho_\star V_\star r_\star^2 , \qquad V_M = \left(\frac{\Omega^2 \mathcal{F}_B^2}{\mathcal{F}_m} \right)^{1/3} . \qquad (26)$$

The speed V_M is a characteristic speed introduced by Michel [42], and plays a central role in magnetic winds. If $V_r \ll V_M$, magnetic effects can be ignored while when $V_r \gg V_M$ they dominate.

For our Sun today,

$$V_M = \left(\frac{\Omega^2 \mathcal{F}_B^2}{\mathcal{F}_m} \right)^{1/3} \simeq 60 - 90 \text{ km/sec} \ll V_{Parker} \approx 600 \text{ km/sec} . \qquad (27)$$

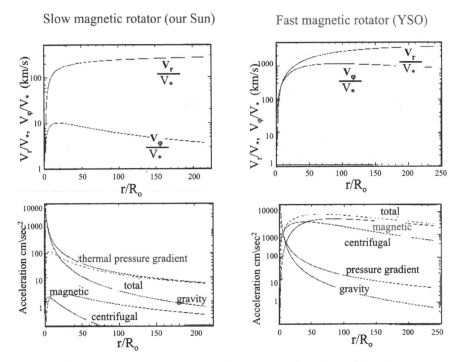

Fig. 11. Upper left: Radial and azimuthal speeds as functions of the dimensionless distance from a WD solution with parameters appropriate to a slow magnetic rotator like our Sun. **Lower left**: Radial accelerations in the above WD solution: total, magnetic, pressure gradient, centrifugal. **Upper/lower right**: the same with the left panels but with parameters appropriate to a fast magnetic rotator like a young star aging 50×10^6 years

Thus, our Sun is a *slow* magnetic rotator (SMR), although in the past it was a *fast* magnetic rotator (FMR). Solutions corresponding to the two extreme cases of a SMR and a FMR are shown in Fig. 11.

4.5 Fast Magnetic Rotators

In the case of fast magnetic rotators, the total specific energy can be calculated asymptotically,

$$\mathcal{E} \approx \Omega L = \Omega^2 r_\star^2 = \frac{1}{2} V_\infty^2 - \left. \frac{\Omega r B_\phi}{\Psi_A} \right|_\infty , \tag{28}$$

In terms of the dimensionless rotational speed w at the Alfvenic distance, $w = (\Omega r_\star)/V_\star$ and the dimensionless radial speed $u(R) = V(r)/V_\star$, the dimensionless Bernoulli integral can be written:

$$2w^2 = (u^\infty)^2 + \frac{2w^2}{u^\infty} . \tag{29}$$

The minimum value of $[u_\infty^2 + 2w^2/u_\infty]$ is equal to $3w^{4/3}$ and corresponds to $u_\infty = w^{2/3}$. Then, we have two solutions u_1^∞, u_2^∞ only when

$$\left. \begin{array}{l} 2w^2 \geq 3w^{4/3} \\[2mm] V_M = V_\star w^{2/3} \end{array} \right\} \Rightarrow V_M \geq \frac{3}{2} V_\star , \text{or}, \ w \geq \left(\frac{3}{2} \right)^{3/2} . \tag{30}$$

For a given value of $\mathcal{E}$, one of these solutions is superfast (u_2^∞) and the other (u_1^∞) subfast. If the value of $\mathcal{E}$ is such that $V_M = (3/2)V_\star$, or, $w = (3/2)^{3/2}$, the fast critical point is at infinity. For the superfast solution:

$$\frac{\mathcal{E}_{Poynting}}{\mathcal{E}_{kinetic}} = 2 \left(\frac{w^{2/3}}{u} \right)^3 < 2 , \tag{31}$$

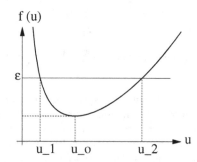

Fig. 12. Plot of the function $f(u) = u^2 + \frac{2w^2}{u}$

because for these solutions $u > \omega^{2/3}$. It follows that $\mathcal{E}_{poynt} < \frac{2}{3}\mathcal{E}$, $\mathcal{E}_{kin} > \frac{1}{3}\mathcal{E}$.

Consider then a Weber/Davis magnetized wind starting with all its available at the base total energy in the Poynting form. Then, if asymptotically the outflow speed equals the fast speed, 1/3 of this available at the base total energy is transformed to kinetic energy of the outflowing plasma, while the other 2/3 remains as Poynting energy.

4.6 Superfast Solutions at Large Distances and Causality

The dispersion relation for the fast Alfvén speed is:

$$V_f^4 - V_f^2(V_s^2 + V_A^2) + V_{Ap}^2 V_s^2 = 0 , \qquad (32)$$

where V_A is the total Alfvén speed, V_{Ap} is the poloidal component of the Alfvén speed and V_s is the sound speed,

$$V_A^2 = \frac{B_r^2 + B_\phi^2}{4\pi\rho} , \qquad V_{Ap}^2 = \frac{B_p^2}{4\pi\rho} = \frac{B_r^2}{4\pi\rho} . \qquad (33)$$

As $R \to \infty$, some solutions have $\rho^\infty \to 0$ and $V_s = \gamma\rho^{\gamma-1} \to 0$. Thus, asymptotically as $V_s \to 0$ in the above dispersion relation, the fast Alfvén speed becomes:

$$V_f^2 = V_A^2 = \frac{B_r^2 + B_\phi^2}{4\pi\rho} . \qquad (34)$$

At the same time, from magnetic flux conservation we have

$$B_r^\infty = \frac{B_\star}{R^2} \to 0 , \qquad (35)$$

while the azimuthal component B_ϕ dominates in all these large distances:

$$B_\phi = \frac{\Omega r_\star \Psi_A}{R} \frac{1 - R^2}{M^2 - 1} \simeq -\frac{\Omega r_\star \Psi_A R}{M^2} . \qquad (36)$$

Hence,

$$V_f^2 \simeq \frac{B_\phi^2}{4\pi\rho} = \frac{\Omega^2 r_\star^2}{M^2/R^2} = \frac{\Omega^2 r_\star^2}{V_\star^2} \frac{V_\star^2}{u} = \frac{\omega^2 V_\star^2}{u} . \qquad (37)$$

Finally, the fast Alfvén number takes the form:

$$M_f^2 = \frac{V_r^2}{V_f^2} = \frac{V_r^2 u}{\omega^2 V_\star^2} = \frac{u^3}{\omega^2} = \left(\frac{u}{\omega^{2/3}}\right)^3 . \qquad (38)$$

Evidently for the solution, $u_2^\infty > \omega^{2/3}$ it holds $M_f > 1$, while for the solution $u_1^\infty < \omega^{2/3}$ we have $M_f < 1$.

In conclusion, only when the flow speed at infinity is larger than the Michel speed it is possible to have superfast speeds at infinity. The existence of this type of superfast solutions is required from the causality principle as well, since perturbations at infinity cannot (and should not) influence the conditions at the base, when the flow speed exceeds the fast speed. The situation is analogous to the case wherein the terminal speed of a Parker HD wind is supersonic and perturbations at infinity cannot influence the conditions at the base of the wind, since the flow speed exceeds the sound speed.

4.7 Angular Momentum Loss via Magnetized Winds

A direct application of the above theory is on the observed spin down of stars in young clusters, wherein as the stars age their rotational speed statistically declines, Fig. 13.

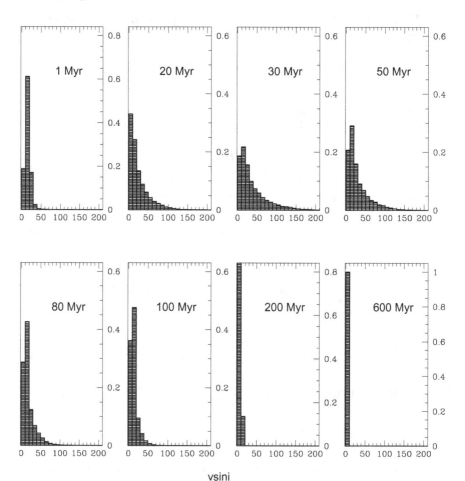

vsini

Fig. 13. Evolution of stellar rotation as evidenced from a synthetic distribution of $V_\phi sini$ in stellar clusters of various ages [12]

Consider then for a moment the angular momentum loss from a shell of the wind of radius R rotating with angular velocity Ω. Since $dJ_{sh} = dI_{sh}\Omega$, the angular momentum loss rate is

$$dI_{sh} = \frac{2}{3}(dM)R^2 \implies dJ_{sh} = \frac{2}{3}R^2\Omega\, dM \implies \frac{dJ_{sh}}{dt} = \frac{2}{3}R^2\Omega\,\dot{M}\ . \quad (39)$$

Application to the Sun: Let τ_J^o the angular momentum loss time scale without a magnetic field and τ_J^M the angular momentum loss time scale with a magnetic field.

1. Angular momentum loss time scale for a young sun without a magnetized wind:

$$\tau_J^o \simeq -\frac{J}{\frac{dJ}{dt}} = \frac{I\Omega}{(2/3)\Omega R_\odot^2\,\dot{M}} = \frac{3kM}{2\dot{M}}\ ,\quad I = kMR_\odot^2\ , \quad (40)$$

for $\dot{M} \simeq 10^{-14} M_\odot$ yr^{-1}, $k \simeq 6 \times 10^{-2}$, since the sun has most of its mass concentrated at its center. Then,

$$\tau_J^o \simeq 600 \times 10^{10} \text{ yrs} \gg \tau_\odot \approx 10^{10} \text{ yrs}\ , \quad (41)$$

where $\tau_\odot$ is the solar age. Thus the removal of angular momentum by a nonmagnetized wind is inefficient.

2. Angular momentum loss time scale with a magnetic field:

$$\tau_J^M \simeq -\frac{J}{\frac{dJ}{dt}} = \frac{I\Omega}{(2/3)\Omega r_A^2\,\dot{M}} = \frac{3kMR_\odot^2\Omega}{2\lambda^2 R_\odot^2\,\dot{M}} = \frac{\tau_J^o}{\lambda^2}\ , \quad (42)$$

with $r_A = \lambda R_\odot$, $\lambda \simeq (10 - 50)R_\odot$. In this case,

$$\tau_J \simeq (0.24 - 6) \times 10^{10} \text{ yrs} \sim \tau_\odot \approx 10^{10} \text{ yrs}\ . \quad (43)$$

Thus, the removal of angular momentum by a magnetized wind is efficient. Figure 13 shows the evolution of stellar rotation as evidenced from synthetic distribution of $V \sin i$ in stellar clusters of various ages. Note that a considerable percentage of stars in young clusters aging several million years rotate rapidly because of contraction and angular momentum conservation. On the other hand, in old clusters aging hundreds of millions of years, most stars have slowed down because of magnetic braking by WD-type magnetized winds.

4.8 Skumanich's Law

Observations of coronal X-ray luminosity give that the X-ray luminosity is proportional to the square of the angular velocity, $L_X \propto \Omega^2$. On the other hand, from the theory of magnetic coronal heating, the X-ray luminosity is proportional to the square of the magnetic field, $L_X \propto B^2$. Then, the stellar magnetic field B_0 is proportional to the rotation rate Ω, $B_o = \chi\Omega$, in accordance and with the Dynamo mechanism. With an angular momentum loss via a magnetized wind we have:

$$-\frac{dJ}{dt} = \frac{2}{3}\Omega r_A^2 \dot{M} \,, \tag{44}$$

For a radial outflow $\dot{M} = 4\pi\rho_A V_A r_A^2$, with $V_A^2 = B_A^2/4\pi\rho_A$, $r_A^4 B_A^2 = r_0^4 B_0^2$,

$$-\frac{dJ}{dt} = \frac{2}{3}\frac{\Omega r_A^4}{V_A}4\pi\rho_A V_A^2 = \frac{2}{3}\frac{\Omega r_A^4 B_A^2}{V_A} = \frac{2}{3}\frac{\Omega B_o^2 r_o^4}{V_A} = \underbrace{\frac{2}{3}\frac{\chi^2 r_o^4}{V_A}}_{\kappa}\Omega^3 = \kappa\Omega^3 \,. \tag{45}$$

But,

$$\frac{dJ}{dt} = I\dot{\Omega} \implies -\dot{\Omega} = \kappa\Omega^3 \implies \Omega(t) = \frac{\Omega_o}{\sqrt{1 + 2\kappa\Omega_o^2 t}} \,. \tag{46}$$

For long time intervals t this simplifies to:

$$\Omega(t) = \frac{1}{\sqrt{2\kappa}}t^{-0.5} \,. \tag{47}$$

which is the verified observationally Skumanich's law [62].

5 Simple Physics in MHD Disk-winds

MHD disk winds offer a nice opportunity to illustrate the basic physics of outflow acceleration, a criterium for jet launching together with a simplified picture for the angular momentum removal from the disk by a jet.

5.1 Minimum Fieldline Inclination for Jet Acceleration

In the following we briefly outline a simple geometrical criterium for the initial minimum fieldline inclination on the poloidal plane for outflow acceleration in disk winds (Fig. 14). We consider on the disk plane (z=0) a unit of plasma volume. At the inner parts of the accretion disk, the Keplerian rotation has kinetic energy greater than the corresponding magnetic energy. On the other hand, on the poloidal meridional plane $r - z$, the magnetic field is sufficiently strong such as it guides the charged particles in Larmor motions around the magnetic field lines with their guiding center sliding along these lines,

$$\frac{1}{2}\rho V_\phi^2 >> \frac{B_p^2}{8\pi} >> \frac{1}{2}\rho V_p^2 \,, \tag{48}$$

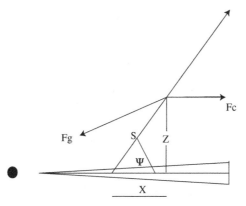

Fig. 14. Illustration for the calculation of the minimum inclination angle of the magnetic fieldlines at the level of the disk for magnetocentrifugal jet acceleration. F_g denotes the central gravitational force while F_C the centrifugal force

i.e. close to the disk, the Keplerian rotational speed is much greater of the Alfvén speed which in turn is much higher than the initial poloidal speed. The basic mechanical analogue is beads (the charged particles) around rigid wires (the poloidal magnetic field lines) while these wires are rotating around the symmetry z-axis with the Keplerian velocity,

$$\Omega_K = \left(\frac{GM}{\varpi_o^3} \right)^{1/2} .$$

The centrifugal force on the rotating charged particles derives from a corresponding centrifugal potential such that the total potential that the particles see is:

$$\Phi(x, z) = -\frac{GM}{\varpi_o} \left\{ \frac{1}{2} \left(\frac{\varpi_o + x}{\varpi_o} \right)^2 + \frac{\varpi_o}{[(\varpi_o + x)^2 + z^2]^{1/2}} \right\} =$$

$$-\frac{GM}{\varpi_o} \left\{ \frac{1}{2} + \frac{x}{\varpi_o} + \frac{x^2}{2\varpi_o^2} + 1 - \frac{x}{\varpi_o} - \frac{x^2}{2\varpi_o^2} - \frac{z^2}{2\varpi_o^2} + \frac{3}{2} \frac{x^2}{\varpi_o^2} + \ldots \right\} =$$

$$-\frac{GM}{\varpi_o} \left\{ \frac{3}{2} + \frac{3x^2 - z^2}{2\varpi_o^2} \right\} , \qquad (49)$$

where we write the cylindrical distance as $\varpi = \varpi_0 + x$ and consider plasma particles close to the disk such that for $x \ll \varpi_0$, $z \ll \varpi_0$ we may employ a Taylor expansion.

Denote by s the arc length on the poloidal fieldline measured from the base of the outflow at the disk level $z = 0$. Also, denote by ψ the angle this poloidal fieldline makes with the disk level, $z = s \sin \psi$; $x = s \cos \psi$. Substituting we get the total potential as a function of s and ψ:

$$\Phi(s, \psi) = -\frac{GM}{\varpi_o} \left\{ \frac{3}{2} + s^2 \frac{3\cos^2 \psi - \sin^2 \psi}{2\varpi_o^2} \right\} . \qquad (50)$$

The total force F_s acting on the plasma particles at the distance s at a given angle ψ is:

$$F_s = -\left.\frac{\partial \Phi}{\partial s}\right|_\psi = \frac{GMs}{\varpi_o^3} \left\{ 3\cos^2 \psi - \sin^2 \psi \right\} . \qquad (51)$$

Therefore, when $F_s > 0$ we have acceleration and plasma outflow from the disk,

$$F_s > 0 \implies 3\cos^2 \psi > \sin^2 \psi \implies \tan^2 \psi < \frac{1}{3} \implies \psi < 60° . \qquad (52)$$

It thus follows, that the fieldlines need to be inclined to the disk midplane by an angle less than 60 degrees for an outwards plasma acceleration. This is usually referred to as the "bead on a rotating wire" analogy.

5.2 Basics of Jet Acceleration

Figure 15 briefly illustrates how an outflow is magnetocentrifugally accelerated up to the Alfvén distance, after which the azimuthal magnetic field eventually collimates the outflow via the built up magnetic hoop stress. The process of acceleration/collimation can be analysed in three steps. *First*, on the disk level the magnetic field is carried around the central object by the rapid

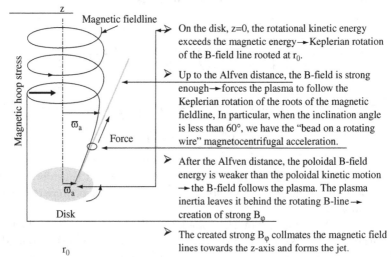

Basics of jet acceleration and collimation

On the disk, z=0, the rotational kinetic energy exceeds the magnetic energy → Keplerian rotation of the B-field line rooted at r_0.

Up to the Alfven distance, the B-field is strong enough → forces the plasma to follow the Keplerian rotation of the roots of the magnetic fieldline, In particular, when the inclination angle is less than 60°, we have the "bead on a rotating wire" magnetocentrifugal acceleration.

After the Alfven distance, the poloidal B-field energy is weaker than the poloidal kinetic motion → the B-field follows the plasma. The plasma inertia leaves it behind the rotating B-line → creation of strong B_φ

The created strong B_φ collimates the magnetic field lines towards the z-axis and forms the jet.

Fig. 15. Illustration of the jet acceleration in radially self-similar outflows

azimuthal rotation. *Second*, just above the disk-plane, the magnetic energy of the poloidal magnetic field exceeds the kinetic energy of the poloidal outflow and thus the plasma moves along the magnetic field on the poloidal plane, with the magnetic field enforcing corotation (as a "bead on a rigid axial wire"). And *third*, after the Alfvén distance, the plasma has gained enough energy to move independently of the magnetic field ("as a rapidly moving metallic bead moves independently of its thin axial plastic filament"). As a result, the rotating rigidly magnetic field winds up, strong azimuthal field is created and the magnetic hoop stress finally collimates the outflow. The Lorenz force can be written,

$$\frac{\boldsymbol{J} \times \boldsymbol{B}}{4\pi} = -\boldsymbol{\nabla} \left(\frac{B^2}{8\pi} \right) + \frac{(\boldsymbol{B} \cdot \boldsymbol{\nabla})\boldsymbol{B}}{4\pi} = \boldsymbol{\nabla} P_m + \boldsymbol{T}_m , \tag{53}$$

where P_m is the magnetic pressure and $\boldsymbol{T}_m$ the magnetic tension. For an axisymmetric azimuthal magnetic field B_ϕ, the magnetic tension is,

$$\boldsymbol{T}_m = \frac{(\boldsymbol{B} \cdot \boldsymbol{\nabla})\boldsymbol{B}}{4\pi} = -\frac{B_\phi^2}{4\pi} \kappa \hat{\boldsymbol{n}} = -\frac{B_\phi^2}{4\pi\varpi} \hat{\boldsymbol{n}} , \quad \frac{d\hat{\phi}}{ds} = \kappa \hat{\boldsymbol{n}} . \tag{54}$$

and it is directed towards $\hat{\boldsymbol{n}}$ which points to the center of curvature of the circular line of radius ϖ.

In other words, one may simulate the magnetic field up to the Alfvén distance to a rigid wire and after to a thin plastic filament with the rotating rigid wires enforcing the winding of the filaments further downstream.

5.3 Transfer of Disk Angular Momentum

In order that accretion on the central object finally occurs, some part of the angular momentum of the rotating plasma in the inner parts of the disk has to be removed. If the friction of any adjacent rotating plasma rings is sufficient, i.e. if the viscosity coefficient is large enough, this angular momentum is transferred to the external parts of the disk by the friction. However, in astrophysical disks the kinematic viscosity is rather low and thus it cannot play such a role. Another possibility is the development of turbulence, for example via the magnetorotational instability.

On the other hand, an efficient way to remove the excess disk angular momentum is provided by a magnetized outflow. A Keplerian disk rotating with angular frequency Ω_K and accreting at a rate $\dot{M}_a$ carries an angular momentum J_a. If this angular momentum is removed at a rate $\dot{J}_a$ from a disk radius ϖ_o, we have:

$$\dot{J}_a = \frac{1}{2} \Omega_K \varpi_o^2 \dot{M}_a . \tag{55}$$

On the other hand, a disk-wind outflowing at a rate $\dot{M}_w$ carries away angular momentum J_w with a rate:

$$\dot{J}_w = \Omega_K \varpi_A^2 \dot{M}_w \ . \tag{56}$$

If the disk-wind carries away a fraction f $(0 < f < 1)$ of the angular momentum of the accreting matter, $\dot{J}_w = f \dot{J}_a$, then,

$$\frac{\dot{M}_w}{\dot{M}_a} = \frac{f}{2} \frac{\varpi_o^2}{\varpi_A^2} \ . \tag{57}$$

With a magnetic lever arm $\varpi_A \sim (5 - 10)\varpi_o$, the disk-wind needs then to carry away only a few percent of the accreting mass rate. This is a rather remarkable fact, in the sense that a small fraction of the outflowing gas can take away most of the angular momentum of the disk allowing thus the disk material to accrete onto the central condensation and form the young star.

Recent observations have shown signatures of jet rotation in the form of transverse velocity shifts of the order of 10 km/sec detected in 6 microjets on scales of z=50 AU, [1, 80]. In particular, out of 4 disks investigated, 2 rotate in the same sense as their jets (DG Tau, CW Tau), 1 is undetermined (HH30) and 1 opposite (RW Aur) [16]. It is interesting to note in passing that some MHD disk-wind models do show such reversal of rotation, ([74], Fig. 5). Also, by using disk wind theory these observations for the case of RW Aur are consistent with the ratio B_φ/B_p of the toroidal and poloidal components of the magnetic field at the observed location (i.e. about 80–100 AU above the disk) to be 3.8 ± 1.1 at 30 AU from the axis in the red lobe and −8.9 ± 2.7 at 20 AU from the axis in the blue lobe (assuming cylindrical coordinates centered on the star), [80]. These observations suggest that the jet should be magnetically collimated, the toroidal component dominates and thus the angular momentum may be removed magnetically from the disk via the jet [24].

6 Steady-state, Exact Self-similar MHD Solutions

As discussed in Sect. 2, magnetized outflows are described to first order by the ideal MHD equations, i.e. Maxwell's equations combined with the conservation of mass, momentum, and energy. Assuming steady-state and axisymmetry, several conserved quantities along the flow exist, [65]. If we label each poloidal field line with the poloidal magnetic flux function A, they are the mass-to-magnetic flux ratio,

$$A = \frac{1}{2\pi} \int \mathbf{B}_p \cdot d\mathbf{S}, \ \ \Psi_A(A) = \frac{4\pi\rho V_p}{B_p} \tag{58}$$

the field angular velocity $\Omega(A)$ and the specific total angular momentum $L(A)$

$$\Omega(A) = (V_\varphi/\varpi) - (V_p/\varpi)(B_\varphi/B_p), \ \ L(A) = \varpi V_\varphi - \varpi B_\varphi/\Psi_A \ , \tag{59}$$

while the Alfvénic lever arm ϖ_a on each field line is, $\varpi_a = \sqrt{L(A)\Omega(A)}$, [65]. It is convenient to introduce two more dimensionless functions, the Alfvén Mach number M and the cylindrical distance in units of ϖ_a,

$$M = \frac{V_p}{B_p/\sqrt{4\pi\rho}}, \quad G = \frac{\varpi}{\varpi_a}. \tag{60}$$

All physical quantities can be expressed as functions of the magnetic flux function A [i.e. $\varpi_a(A)$, $\Psi_A(A)$, $\Omega(A)$] and the two variables (G, M) (e.g. see [65])

$$\mathbf{B} = \frac{dA}{d\varpi_a}\nabla\frac{\varpi}{G} \times \frac{\hat{\varphi}}{\varpi} - \frac{\varpi_a^2\Omega\Psi_A}{\varpi}\frac{1 - G^2}{1 - M^2}\hat{\varphi}, \tag{61}$$

$$\mathbf{V} = \frac{M^2}{\Psi_A}\mathbf{B}_p + \frac{\varpi_a^2\Omega}{\varpi}\frac{G^2 - M^2}{1 - M^2}\hat{\varphi}. \tag{62}$$

The functions $G(r, \theta)$ and $M(r, \theta)$ can be obtained by integrating the two coupled components of the momentum equation on the poloidal plane. However, due to the complexity of these two partial differential equations (PDE) and in order to proceed semi-analytically, we are forced to make further assumptions. For example, if we are interested for a nonlinear separation of the variables in the two PDE, we may employ the only known at present approach, namely the method of self-similarity.

A physical phenomenon is called *temporarily* self-similar if it can be reproduced at any time via a self-similar mechanism from a previous temporal state. The classical such example is a nuclear bomb explosion with the mushroom typical shape.

Analogously, a physical phenomenon is called *spatially* self-similar if it can be reproduced everywhere in space via a spatial self-similar mechanism. The classical such example are the Russian matryoshkas. Also, the observed shapes of astrophysical jets in the galactic and extragalactic scales, are suggestive of such a symmetry in space. Technically, spatial self-similarity may be viewed as a method of nonlinear separation of the variables in the set of the steady MHD equations, providing us the opportunity to obtain analytically solutions.

In particular, in self-similarity it is assumed that both G and M are functions of a single variable χ, [74]. If this is the case, the ratios $(1-G^2)/(1-M^2)$, $(G^2 - M^2)/(1 - M^2)$ appearing in (61) and (62) are functions of χ only, and the components of the momentum equation become relatively simple expressions of χ and ϖ_a. It is in principle possible to choose the functional form of the integrals such that the variables (χ, ϖ_a) decouple, in which case the equations become ordinary differential (ODEs) with respect to χ. The only remaining difficulty then is that the solution should cross various singular points, corresponding to ratios $\frac{0}{0}$ in the ODEs.

This unifying scheme contains two large families of exact MHD models, which are systematically constructed in [74]:

1. For $\chi = \theta$ we get the family of the *radially* self-similar models with conical critical surfaces and with prototype the Blandford & Payne model, [4] [see also [76] for the relativistic case]. Figure 16(a) illustrates the radial self-similar character of the poloidal field lines, resulting from the assumption $\varpi = \varpi_a G(\theta)$.

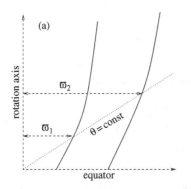 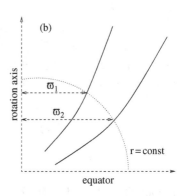

Fig. 16. An illustration of the two possibilities of self-similar field line structure. Consider two arbitrary field lines (*thick lines*). In **(a)** is illustrated the case of radial self-similarity, wherein the ratio ϖ_1/ϖ_2 for the intersection of any poloidal line with a cone is the same for any value of θ. Specifically, if we write $A(r,\theta) = \varpi^x f(\theta)$, the ratio of the cylindrical distances ϖ_1 and ϖ_2 in which the direction θ=const. intersects the two lines $A = A_1$ and $A = A_2$ is $A_1/A_2 = (\varpi_1/\varpi_2)^x$. Thus, if the MHD problem is solved for one fieldline $A = A_1$, i.e. the unknown function $f(\theta)$ is calculated, from the distance ϖ_1 we may calculate ϖ_2, $\varpi_2 = A_2/A_1^{1/x}\varpi_1$ for given values of $A = A_1$ and $A = A_2$. Thus, if we know one field line we may construct all the others. In **(b)** it is illustrated the case of meridional self-similarity. By writing $A(r,\theta) = \varpi^x f(r)$, a determination of the unknown function $f(r)$ yields the ratio of the cylindrical distances ϖ_1 and ϖ_2 in which the sphere r=const. intersects the two lines $A = A_1$ and $A = A_2$. This ratio is the same for any spherical surface r =const., i.e. $\varpi_2 = (A_2/A_1)^{1/2}\varpi_1$. Thus, again if we know one field line we may construct all the others

2. For $\chi = r$ we get the family of the *meridionally* self-similar models with spherical critical surfaces and with prototype the Sauty & Tsinganos model [55](henceforth ST94 model; see also [58]). This family also includes the classical Parker [50] description of a stellar wind, as its simplest member; it also contains the simple prescribed field line model of Tsinganos & Trussoni, [67]. Figure 16(b) illustrates the meaning of the meridional self-similar assumption $\varpi = \varpi_a G(r)$.

In Table 1 we summarize all possible classes of the meridionally self-similar family of solutions (for details see [74]). Instead of using the three functions of the dimensionless poloidal flux function α, $(\mathcal{A}, \Psi_A, \Omega)$, we found it more convenient to use the three dimensionless functions $g_1(\alpha)$, $g_2(\alpha)$ and $g_3(\alpha)$,

$$g_1(\alpha) = \int \mathcal{A}'^2 d\alpha \,, \ g_2(\alpha) = \frac{r_\star^2}{B_\star^2} \int \Omega^2 \Psi_A^2 d\alpha \,, \ g_3(\alpha) = \frac{\Psi_A^2}{4\pi\rho_\star} \,. \tag{63}$$

Similarly, in the corresponding Table 2 which summarizes all possible classes of the radially self-similar family of solutions, we found it more convenient to use the three dimensionless functions $q_1(\alpha)$, $q_2(\alpha)$, $q_3(\alpha)$ (for details see [74]),

Table 1. Meridionally Self-similar Models

Case	$g_1(\alpha)$	$g_2(\alpha)$	$g_3(\alpha)$
(1)	α	$\lambda^2\alpha$	$1+\delta\alpha$
(2)	α	$\xi\alpha+\mu\alpha^\epsilon/\epsilon$	$1+\delta\alpha+\mu\delta_0\alpha^\epsilon$
(3)	α	$\xi\alpha+\mu\alpha\ln\alpha$	$1+\delta\alpha+\mu\delta_0\alpha\ln\alpha$
(4)	$\alpha_0 e^{\frac{\alpha}{\alpha_0}}$	$\lambda e^{\frac{\alpha}{\alpha_0}}$	$1+\delta\alpha e^{\frac{\alpha}{\alpha_0}}+\mu\left(e^{\frac{\alpha}{\alpha_0}}-1\right)$
(5)	$\dfrac{\alpha_0}{\epsilon}\mid\dfrac{\alpha}{\alpha_0}-1\mid^\epsilon$	$\xi\mid\dfrac{\alpha}{\alpha_0}-1\mid^\epsilon$	$1+\delta\mid\dfrac{\alpha}{\alpha_0}-1\mid^\epsilon+\mu\mid\dfrac{\alpha}{\alpha_0}-1\mid^{\epsilon-1}-\delta-\mu$
(6)	$-\alpha_0\ln\mid\dfrac{\alpha}{\alpha_0}-1\mid$	$\xi\ln\mid\dfrac{\alpha}{\alpha_0}-1\mid$	$1+\delta\ln\mid\dfrac{\alpha}{\alpha_0}-1\mid+\mu\dfrac{\alpha}{\alpha_0(\alpha-\alpha_0)}$
(7)	$\dfrac{\alpha}{1-\alpha_0}$	$\mu\ln\dfrac{\alpha}{\alpha_0}+\xi\alpha$	$1+\delta(\alpha-\alpha_0)+\mu\delta_0\ln\dfrac{\alpha}{\alpha_0}$
(8)	$\dfrac{\alpha_0}{\epsilon(1-\alpha_0)}\left(\dfrac{\alpha}{\alpha_0}\right)^\epsilon$	$\lambda_1\alpha^\epsilon+\lambda_2\alpha^{\epsilon-1}$	$1+\delta_1(\alpha^\epsilon-\alpha_0^\epsilon)+\delta_2\left(\alpha^{\epsilon-1}-\alpha_0^{\epsilon-1}\right)$
(9)	$\dfrac{\alpha_0}{1-\alpha_0}\ln\dfrac{\alpha}{\alpha_0}$	$\lambda_1\ln\dfrac{\alpha}{\alpha_0}+\dfrac{\lambda_2}{\alpha}$	$1+\delta_1\ln\dfrac{\alpha}{\alpha_0}+\delta_2\left(\dfrac{1}{\alpha}-\dfrac{1}{\alpha_0}\right)$

Table 2. Radially Self-similar Models

Case	$q_1(\alpha)$	$q_2(\alpha)$	$q_3(\alpha)$
(1)	$\dfrac{E_1}{F-2}\alpha^{F-2}$	$\dfrac{D_1}{F-2}\alpha^{F-2}$	$\dfrac{C_1}{F-2}\alpha^{F-2}$
(2)	$E_1\ln\alpha$	$D_1\ln\alpha$	$C_1\ln\alpha$
(3)	$E_1\alpha^{x_1}+E_2\alpha^{x_2}$	$D_1\alpha^{x_1}+D_2\alpha^{x_2}$	$C_1\alpha^{x_1}+C_2\alpha^{x_2}$
(4)	$E_1\ln\alpha+E_2\alpha^x$	$D_1\ln\alpha+D_2\alpha^x$	$C_1\ln\alpha+C_2\alpha^x$
(5)	$E_1(\ln\alpha)^2+E_2\ln\alpha$	$D_1(\ln\alpha)^2+D_2\ln\alpha$	$C_1(\ln\alpha)^2+C_2\ln\alpha$
(6)	$E_1\alpha^x\ln\alpha+E_2\alpha^x$	$D_1\alpha^x\ln\alpha+D_2\alpha^x$	$C_1\alpha^x\ln\alpha+C_2\alpha^x$

$$q_1(\alpha)=\int\frac{A'^2}{\alpha}d\alpha\;,\;q_2(\alpha)=\frac{\varpi_o^2}{B_o^2}\int\Omega^2\Psi_A^2 d\alpha\;,\;q_3(\alpha)=\frac{\mathcal{GM}}{B_o^2\varpi_o}\int\frac{\Psi_A^2}{\alpha^{\frac{3}{2}}}d\alpha \quad (64)$$

It is worth to note that in the class of meridionally self-similar solutions analyzed in [55] an interesting parametric energetic criterion emerges which characterizes the asymptotic shape of the streamlines. In terms of this parameter, we may either have an Efficient Magnetic Rotator magnetic rotator, efficient (**EMR**) to magnetically collimate the outflow into a jet, or, an Inefficient Magnetic Rotator (**IMR**). This EMR/IMR criterion is an extension to 2-D of the FMR/SMR criterion in the 1-D Weber & Davis model. For more details and application of this criterion to the various astrophysical outflows, the interested reader is referred to the Lecture of C. Sauty in this volume.

6.1 Critical Points, Separatrices and Causality

An interesting feature of axisymmetric MHD wind-type solutions is the appearance of two X-type critical points within the flow domain, in addition to the Alfvén critical point. In general, at the critical points the bulk flow speed equals to one of the characteristic speeds in the problem. Hence, it is of physical interest to associate the flow speeds at these X-type critical points to a characteristic speed for MHD wave propagation. In that connection, first note that these semi-analytical solutions possess the symmetries of self-similarity and axial symmetry. Thus, in spherical coordinates (r, θ, φ), the self-similarity direction $\hat{s}$ can be $\hat{s} \equiv \hat{\theta}$, or, $\hat{s} \equiv \hat{r}$ and the axisymmetry direction is $\hat{\varphi}$. Therefore, a wave that preserves those two symmetries should propagate along the $\hat{\varphi} \times \hat{s} \parallel \hat{\chi}$-direction in the meridional plane. Besides the incompressible Alfvén mode propagating along the magnetic field ($\mathbf{B}$) with velocity V_A, the compressible slow/fast MHD modes propagate in the direction $\hat{\chi}$ with phase speeds $V_\chi \equiv V_{slow,\chi}$, or, $V_\chi \equiv V_{fast,\chi}$ which satisfy the quartic

$$V_\chi^4 - V_\chi^2(V_A^2 + C_s^2) + C_s^2 V_{A,\chi}^2 = 0 . \tag{65}$$

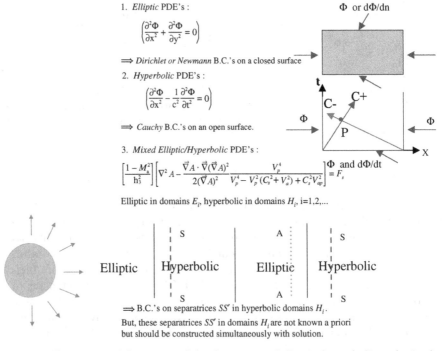

Nature of PDE and Correct Boundary Conditions

1. *Elliptic* PDE's :

$$\left(\frac{\partial^2 \Phi}{\partial x^2} + \frac{\partial^2 \Phi}{\partial y^2} = 0\right)$$

$\Rightarrow$ *Dirichlet or Neumann* B.C.'s on a closed surface

2. *Hyperbolic* PDE's :

$$\left(\frac{\partial^2 \Phi}{\partial x^2} - \frac{1}{c^2}\frac{\partial^2 \Phi}{\partial t^2} = 0\right)$$

$\Rightarrow$ *Cauchy* B.C.'s on an open surface.

3. *Mixed Elliptic/Hyperbolic* PDE's :

$$\left[\frac{1 - M_A^2}{h_3^2}\right]\left[\nabla^2 A - \frac{\vec{\nabla}A \cdot \vec{\nabla}(\vec{\nabla}A)^2}{2(\vec{\nabla}A)^2} \frac{V_p^4}{V_p^4 - V_p^2(C_s^2 + V_a^2) + C_s^2 V_{ap}^2}\right] = F_s$$

Elliptic in domains E_i, hyperbolic in domains H_i, i=1,2,...

$\Rightarrow$ B.C.'s on separatrices SS' in hyperbolic domains H_i.

But, these separatrices SS' in domains H_i are not known a priori but should be constructed simultaneously with solution.

Fig. 17. Illustration of the nature of the three cases of elliptic, hyperbolic and mixed-type partial differential equations and the corresponding appropriate boundary conditions

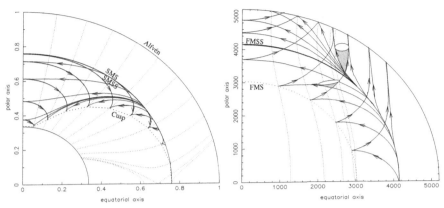

Fig. 18. Slopes of the two families of characteristics for a meridionally self-similar solution in each of the two hyperbolic regimes of the problem from Sauty et al. (2004). In the left figure the slow magnetoacoustic separatrix surface (SMSS) is at $R = r/r_\alpha = 0.751$ just before the slow magnetoacoustic surface (SMS). In the right figure the fast magnetoacoustic separatrix surface (FMSS) is at $R = 4158$ above the fast magnetoacoustic surface (FMS) at about $R = 3000$. Arrows indicate the direction of MHD signal propagation while two Mach cones above and below the FMSS are also shown

Hence, when the above equation is satisfied the governing equations have X-type singularities and $V_\chi = \mathbf{V} \cdot \hat{\chi}$.

On the other hand, it is well known that in the MHD flow system there exist two hyperbolic regimes wherein characteristics exist: the inner hyperbolic regime which is bounded by the cusp and the slow magnetosonic surfaces and the outer hyperbolic regime extending downstream of the fast magnetosonic point, Fig. 17. Within each of those two hyperbolic regimes, there exists one limiting characteristic or separatrix surface: the slow magneto-acoustic separatrix surface (SMSS) inside the inner hyperbolic regime and the fast magneto-acoustic separatrix surface (FMSS) inside the outer hyperbolic regime, [6, 68]. The true critical points are found precisely on these two separatrices. Furthermore, the FMSS plays the role of the MHD signal horizon of the problem in the sense that if the poloidal outflow speed exceeds the corresponding speed at the FMSS, then no perturbation can affect the solution upstream of the FMSS. In other words, setting the boundary conditions at the FMSS is a proxy of setting the correct boundary conditions at infinity. As an example, in Fig. 18 corresponding to a meridionally self-similar case analysed in [58], the SMSS is at $R = 0.751$ while the FMSS is located at $R = 4150$. An analogous situation appears in radially self-similar cases with two critical transitions at the SMSS ($M_{ms} = 1$) and FMSS ($M_{mf} = 1$)]. Such an example is shown in Fig. 19, [75].

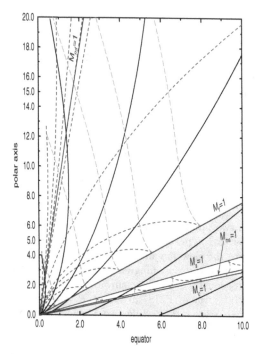

Fig. 19. The two families of characteristics in the two hyperbolic regimes of a radially self-similar solution, [75]. The slow magnetoacoustic separatrix surface (SMSS) is at $\theta \approx 72°$ (indicated by $M_{ms} = 1$) and lies between the cusp surface (indicated by $M_c = 1$) and the slow surface (indicated by $M_s = 1$), the Alfvén surface is at $\theta = 60°$, the fast surface is indicated by $M_f = 1$ and the fast magnetoacoustic separatrix surface (FMSS) is at $\theta \approx 7°$ and is indicated by $M_{mf} = 1$

7 Numerical Simulation of Jet Formation

An investigation of the problem of the collimation of a MHD outflow can be also obtained through a numerical simulation of the time-dependent MHD equations. In one approach we have employed, the simulation can be done in two steps. *First*, a steady state solution in the nearest zone which contains the relevant MHD critical surfaces and the governing PDE are of mixed elliptic/hyperbolic type is obtained by using a relaxation method [10, 70]. In the *second* step, the solution in the far zone can be obtained by extending to large distances the solution obtained in the nearest zone. This ability to extend the inner zone solution is based on the fact that the outflow in the far zone is already superfast magnetosonic. Therefore, the problem can be treated as an initial value Cauchy-type problem with the initial values taken from the solution of the problem in the nearest zone. The advantage of this method is that large lengths of the jet can be modelled. A second method employs direct time-dependent numerical simulation from the base of the outflow, but in this way inevitably a smaller length of the jet is modelled (cf. [18, 36, 47, 72, 81]).

7.1 Why is a Magnetized and Rotating Outflow Collimated?

To illustrate in simple terms the effects of rotation and magnetic fields in the outflow of a plasma from a central gravitating object, consider a monopole-type magnetic field, $B_r = B_o/R^2$, where B_o is the magnetic field at the base $R = r/r_o = 1$. Assume that the plasma flows with a constant speed V_o along these radial magnetic field lines. The Alfvén number of this outflow is,

$$M^2(R) = \frac{V_0^2}{V_A^2} = 4\pi\rho\frac{V_0^2}{B_0^2}R^4 = 4\pi\rho_a\frac{V_0^2}{B_0^2}\frac{R^2}{R_a^2} , \qquad (66)$$

where from mass conservation $\rho = \rho_a(R_a^2/R^2)$. Since at the Alfvén radial distance $R_a : M = 1$, $4\pi\rho_a V_0^2 = B_0^2$ we have finally,

$$M = \frac{R}{R_a} . \qquad (67)$$

Let us assume that the base of the outflow rotates with an angular velocity Ω. From the steady MHD equations the induced azimuthal magnetic field B_φ is

$$\frac{B_\varphi}{B_r} = -\frac{\Omega\varpi_a^2}{\varpi(B_r/\Psi_A)}\frac{\varpi^2/\varpi_a^2 - 1}{M^2 - 1} , \qquad (68)$$

where Ψ_A is the mass flux per unit of magnetic flux. Let us assume for the moment that the ouflow remains radial after the rotation starts. Then, $\varpi = R\sin\theta$ such that $M = \varpi/\varpi_a$ and for distances much larger than the Alfvenic cylindrical distance, $\varpi \gg \varpi_a$, $R \gg R_a$, $M^2 \approx \varpi^2/\varpi_a^2$, it follows that

$$\frac{B_\varphi}{B_r} \approx -\frac{\Omega\varpi_a^2}{\varpi}\frac{\Psi_A}{B_r} = -\frac{\Omega\varpi_a^2}{\varpi}\frac{\Psi_A^2}{4\pi\rho}\frac{4\pi\rho}{\Psi_A B_r} = -\frac{\Omega\varpi_a^2}{\varpi}\frac{\Psi_A^2}{4\pi\rho}\frac{1}{V_0} \qquad (69)$$

But,

$$\frac{\Psi_A^2}{4\pi\rho} = \frac{4\pi\rho V_0^2}{B_r^2} = M^2 \approx \frac{\varpi^2}{\varpi_a^2} , \qquad (70)$$

and thus,

$$\frac{B_\varphi}{B_r} \approx -\frac{\Omega}{V_0}\varpi . \qquad (71)$$

i.e. the azimuthal magnetic field grows with the cylindrical distance ϖ in relation to the poloidal magnetic field B_r. Thus, although at the rotation axis the magnetic tension is negligible, the azimuthal magnetic field grows with distance from the axis of rotation and eventually it will dominate over the poloidal magnetic field B_p. The magnetic pressure and tension then exert a net force towards the axis of rotation and one may wonder for what might balance this inwards force. The outward inertial (centrifugal) force $\rho V_\varphi^2/\varpi$ is negligible since the azimuthal flow speed is negligible in the same approximation,

$$V_\varphi = \frac{\Omega\varpi_a^2}{\varpi}\left[1 - \frac{\varpi^2/\varpi_a^2 - 1}{M^2 - 1}\right] \approx 0 . \tag{72}$$

The last available means to balance the inwards hoop stress would be some suitable pressure gradient. However, the magnetic pressure drops with the cylindrical distance ϖ like $1/\varpi^2$ and is negligible. The thermal gas pressure on the other hand, should drop like $1/\varpi^3$ in an atmosphere where $V = V_o$ ($\rho \sim \varpi^{-2}$) in order that the thermal pressure gradient balances gravity. It follows that the unavoidable result is that magnetic tension will bend the poloidal magnetic field lines towards the axis, forming a cylindrical core. Such

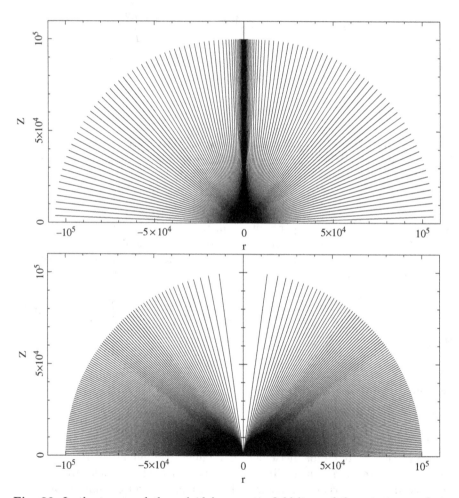

Fig. 20. In the top panel the poloidal magnetic field lines of the rotating outflow are plotted in the far zone and for intervals of equal magnetic flux $\Delta\Phi = 10^{-2}$ for a total normalized flux $\Phi = 1$. For comparison, the original (t=0) nonrotating and uncollimated initial monopole magnetosphere is shown in the bottom panel

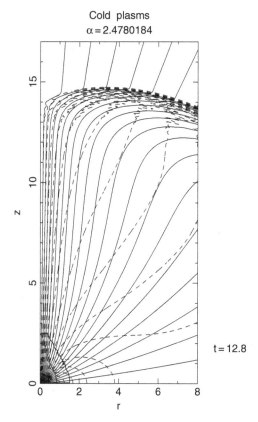

Cold plasms
α = 2.4780184

t = 12.8

Fig. 21. Intermediate state, wherein after the beginning of rotation of the base of the outflow, the magnetic field lines focus towards the axis of rotation, by the Lorentz $j \times B$ force. An MHD shock wave propagates downstream the information of the rotation and collimates at large distances the outflow

a dramatic formation of an inner jet by magnetic self-collimation may be seen in Fig. 20(a) after we start rotating the initial radial magnetosphere of Fig. 20(b): This magnetic confinement is also clearly seen in Fig. 21. In Fig. 22 we rediscover the result of the steady MHD modelling that a fast magnetic rotator (in this case a YSO rotating 10 times faster than the Sun) produces a tightly collimated jet while the solar wind does not show any significant collimation.

7.2 A Two-component Model
for Jets from the System of a Central Source and a Disk

A serious limitation however of the previous simulations of magnetic collimation is that only a tiny fraction of order $\sim 1\%$ of the mass and magnetic flux of the originally radial wind ends up collimated inside the jet [10]. Similarly, in

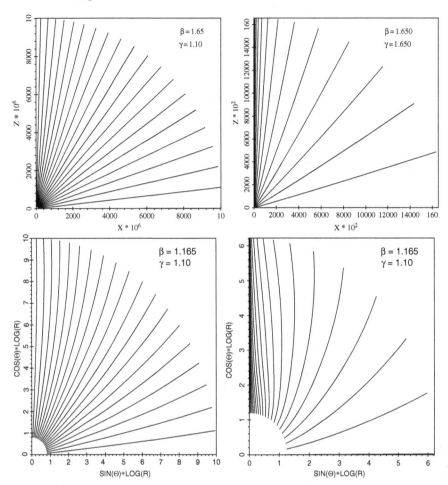

Fig. 22. Left panels: Shape of poloidal magnetic field lines in the far zone of an assumed isotropic solar wind. In the upper panel the poloidal field lines are plotted in a linear scale, while in the lower panel in a logarithmic scale which magnifies their slight bending towards the axis. As it may be seen, collimation is negligible. **Right panels**: Same as in the left panels but for a wind from a star rotating 10 times faster than the Sun. Note the significant collimation in this case

analytical models if the source of the wind is a stellar surface and the disk does not feed the outflow with mass and magnetic flux, very low wind- and jet-mass loss rates ($\dot{M}_{wind}$, $\dot{M}_j$) are obtained. However, in outflows associated with YSO current estimates place $\dot{M}_{jet}$ in the limits $\dot{M}_{jet} \sim 10^{-6} - 10^{-8} M_\odot/yr$ [51]. And, the inferred from observations mass loss rates of bipolar outflows indicate wind mass loss rates also in the range of $\dot{M}_{wind} \sim 10^{-6} - 10^{-8} M_\odot/yr$, depending largely on the luminosity of the YSO's. Therefore, the mass loss rate in the jet has to be a large fraction of the mass loss rate in the surrounding wind. The idea that the source of the jet rotates rather slowly may be quite reasonable, at least in relation to YSO's. It is evident that a protostar should rotate more slowly than the inner edges of its Keplerian accretion disk and observations indeed confirm this prediction [1]. We do not intend to argue here that the matter in the jet is ejected from the protostar. The close disk-jet connection [29] shows that the matter in the jet is supplied by the accretion disk. But it is reasonable to assume that this matter penetrates in the magnetic field of the central star, partially falls down on the surface of the star and partially is ejected outwards, [21, 61]. In this case only the magnetic field of the jet is connected with the central star. Schematically this picture of the outflow is presented in Fig. 23. According to this scheme the disk not only supplies the plasma of the jet, but also it produces the magnetized wind which collimates the outflow from the central source into a jet. In Fig. 24 is shown the asymptotic state wherein an inner radially expanding wind is forced to

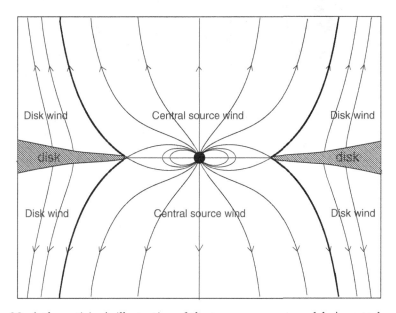

Fig. 23. A theoretician's illustration of the two-component model. A central source emits an initially roughly radially expanding at the base outflow. This stellar wind is self-collimated if the protostar is an EMR. If the protostar is an IMR, the stellar wind is assisted to collimate by the disk wind from the rapidly rotating inner edges of the surrounding accretion disk which is easily self-collimated. Arrows indicate the magnetic field. Typical dimensions for the system of a YSO are : protostellar radius, $R_* = 3R_\odot$, disk inner radius (magnetospheric cavity) $\varpi_i = 0.1AU$, jet emitting part of the disk, 2.5 AU, disk outer radius $\varpi_e = 100AU$. The jet carries most of the angular momentum of the accreting gas which then falls on the protostar along the paths of the magnetic field in the magnetospheric cavity

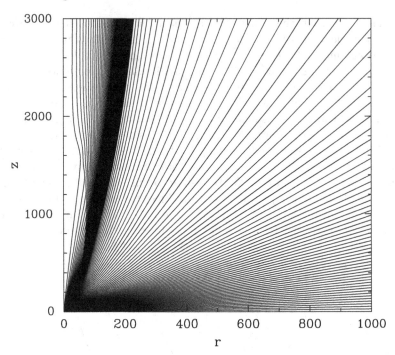

Fig. 24. The result of the simulation for the configuration shown in the previous Fig. (23), [11]. When the disk starts rotating its disk wind collimates forcing the inner wind to collimation too. A byproduct of the collision of the two outflow components is the formation of a shock at the interface of the central wind and the surrounding disk-wind

collimate by the surrounding disk wind. A shock wave is formed at the interface of the two components of the outflow, as it may be seen in Fig. 24, [11].

8 Summary

The purpose of this review lecture was to outline the *unity* of astrophysical MHD outflows, from the Parker classical nonrotating and unmagnetized thermal wind to the 1-D magnetized and rotating Weber & Davis equatorial outflow and finally to the exact 2-D MHD solutions describing winds and jets associated with YSO's. We highlighted some simple physical results of the theory of MHD outflows, for example, the minimum inclination of the fieldlines for plasma acceleration, the mathematical theory of characteristics and their relation with causality and correct definition of the boundary conditions. The method of nonlinear separation of the variables of the governing MHD equations via the self-similarity assumption has been shown to unify all existing analytical solutions for astrophysical outflows, such as the Parker purely thermally driven wind, the thermally or magnetically driven stellar winds which can be thermally or magnetically confined, and finally the purely magnetically driven and magnetically confined disk-winds, [74]. Numerical simulations have shown to reproduce results of the analytical solutions for magnetic self-collimation and a generalization of the fast/slow magnetic rotators theory of the Weber & Davis 1-D modeling to the efficient/inefficient magnetic rotators theory via the existense of an energetic criterion for collimation, [55]. With this parameter, one may understand the appearance of astrophysical outflows in winds and jets which was shown in the first figure of this article, Fig. (1) Finally, we have concluded that the various classes of observed YSO outflows may be understood as a combination of a stellar wind and a disk-wind. Analytical theory provides no room for any other possibility. The precise contribution of each of these two components depends on the stage of evolution of the YSO, with disk-winds dominating in the early phases of star formation and the stellar component left alone in the last ZAMS stage of the star, in the form of the familiar solar/stellar wind, when the disk and its wind have both dispersed. More classes of exact solutions can be examined to test this general theory, together with numerical simulations of MHD outflows, a task that still remains to be pursued.

Acknowledgements

The present work was supported in part by the EU RTN HPRN-CT-2002-00321, the EU Marie Curie Actions Human Resource and Mobility within the JETSET network (JET Simulations, Experiments and Theory) under contract MRTN-CT-2004 005592 and the European Social Fund and National Resources - (EPEAEK II) - PYTHAGORAS. The author thanks Drs S. Bogovalov, C. Sauty, E. Trussoni and N. Vlahakis for their enjoyable collaboration in obtaining some of the results which are reported here.

References

1. Bacciotti, F., 2004, in *Virtual Astrophysical Jets*, Ap&SS, 293, 37
2. Bacciotti, F., Eislöffel, J., 1999, ApJ, 342, 717
3. Biretta, T., 1996, in *Solar and Astrophysical MHD Flows*, K. Tsinganos (ed.), Kluwer Academic Publishers, 357
4. Blandford, R.D. & Payne, D. G., MNRAS, 199, 883 (1982)
5. Bogovalov, S.V., 1992, Sov. Astron. Letts, 18(55), 337
6. Bogovalov, S.V., 1994, MNRAS, 270, 721
7. Bogovalov, S.V., 1996, MNRAS, 280, 39
8. Bogovalov, S.V., 1997, A&A, 327, 662
9. Bogovalov, S.V., Tsinganos, K., 1999, MNRAS, 305, 211 (BT99)
10. Bogovalov, S.V., Tsinganos, K., 2001, MNRAS, 325, 249 (BT01)
11. Bogovalov, S.V., Tsinganos, K., 2005, MNRAS, 357, 918 (BT05)

12. Bouvier, J., Forestini, M., and Allain, S., AA, 1997, 326, 1023
13. Burderi, L., King, A.R., 1995, MNRAS, 276, 1141
14. Burrows, C.J., et al. 1995, ApJ, 452, 680
15. Cabrit, S., 2003, Ap&SS, 287, 259
16. Cabrit, S., Pety, J., Pesenti, N., Dougados, C., 2006, A&A, 452, 897
17. Cao, X., 1997, MNRAS, 291, 145
18. Casse, F., 2004, in *Virtual Astrophysical Jets*, Ap&SS, 293, 91
19. Ciardi, A., Lebedev, S.V., Frank, A., Blackman, E.G., Ampleford, D.J., Jennings, C.A., Chittenden, J.P., Lery, T., Bland, S.N., Bott, S.C., Hall, G.N., Rapley, J., Vidal, F.A., Suzuki, Marocchino, A., 2006, Ap&SS, 307, 17
20. Crane, P., Vernet, J., 1997, ApJ, 486, L91
21. Ferreira, J., & Pelletier, G., 1995, A&A, 295, 807
22. Ferreira, J., 2002, EAS Publications Series, Vol. 3, 229
23. Ferreira, J., 2003, Ap&SS, 287, 15
24. Ferreira, J., Dougados, C., Cabrit, S., 2006, A&A, 453,785
25. Frank, A., Gardiner, T.A., Delamarter, G., Lery, T., Betti, R., 2006, ApJ, 524, 947
26. Gracia, J., Vlahakis, N., Tsinganos, K., 2006, MNRAS, 367(1), 201
27. Goodson, A.P., Winglee, R.M., Bohm, K.H., 1997, ApJ, 489, 199
28. de Gouveia dal Pino, E.M., 2005, Adv. in Sp. Res., 35(5), 908
29. Hartigan, P., Edwards, S., & Ghandour, L., 1995, ApJ, 452, 736
30. Heyvaerts, J., & Norman, C.A., 1989, ApJ, 347, 1055
31. Heyvaerts, J., & Norman, C.A., 2003, ApJ, 596, 1270
32. Jones, D.L., Wehrle, A.E., 1994, ApJ, 427, 221
33. Kafatos, M., 1996, in *Solar and Astrophysical MHD Flows*, K. Tsinganos (ed.), Kluwer Academic Publishers, 585
34. Kahabka, P., Trumper, J., 1996, in *Compact Stars in Binaries*, E.P.J. Van den Heuvel & J. van Paradijs (eds.), (Dordrecht: Kluwer), p. 425
35. Kersale, E., Longaretti, P.-Y., Pelletier, G., 2000, A&A, 363, 1166
36. Krasnopolsky, R., Li, Z.-Y., & Blandford, R., 2000, ApJ, 526, 631
37. Lebedev, S.V., Ciardi, A., Ampleford, D.J., Bland, S.N., Bott, S.C., Chittenden, J.P., Hall, G.N., Rapley, J., Jennings, C.A., Frank, A., Blackman, E.G., Lery, T., 2005, MNRAS, 361(1), 97
38. Lery, T., Henriksen, R.N., Fiege, J.D., Ray, T.P., Frank, A., Bacciotti, F., 2002, A&A, 387, 187
39. Livio, M., 1997, in *13th North American Workshop on Cataclysmic Variables*, 14–19 June 1997, Jackson Hole, WI, USA
40. Low, B.C., Tsinganos, K., 1986, ApJ, 302, 163
41. Massaglia, S., Bodo, G., Rossi, P., S., 2004, Ap&SS, 293, vii
42. Michel, F.C., 1969, ApJ, 158, 727
43. Mirabel, I.F., Rodriguez, L.F., 1996, in *Solar and Astrophysical MHD Flows*, K. Tsinganos (ed.), Kluwer Academic Publishers, 683
44. Nerney, S.F., Swess, J., 1975, ApJ, 196, 837
45. O' Dell, C.R. & Wen, Z., 1994, ApJ, 436, 194
46. Ostriker, E.C., 1997, ApJ, 486, 306
47. Ouyed, R., & Pudritz, R.E., 1997, ApJ, 482, 712
48. Ouyed, R., & Pudritz, R.E., 1997, ApJ, 484, 794
49. Parker, E.N., 1958, ApJ, 128, 664

50. Parker, E.N., 1963, Interplanetary Dynamical Processes, Interscience Publishers, New York
51. Ray, T.P., 1996, in *Solar and Astrophysical MHD Flows*, K. Tsinganos (ed.), Kluwer Academic Publishers, 539
52. Reipurth, B., & Bally, J., 2001, Ann. Rev. Astron. Astroph., 39, 403
53. Sakurai, T., 1985, A&A, 152, 121
54. Sakurai, T., 1987, PASJ, 39, 821
55. Sauty, C., & Tsinganos, K., 1994, A&A, 287, 893 (ST94)
56. Sauty, C., Tsinganos, K., and Trussoni, E., 1999, A&A, 348, 327
57. Sauty, C., Trussoni, E., & Tsinganos, K., 2002, A&A, 389, 1068
58. Sauty, C., Trussoni, E. & Tsinganos, K., 2004, A&A, 421, 797
59. Sauty, C., Lima, J.J.G., Iro, N., & Tsinganos, K., 2005, A&A, 432, 687
60. Shahbaz, T., Livio, M., Southwell, K.A., Charles, P.A., 1997, ApJ, 484, L59
61. Shu, F., Ruden, S.P., Lada, C.J., Lizano, S., 1991, ApJ, 370, L31
62. Skumanich, A., 1972, ApJ., 171, 565
63. Suess, J., 1972, J. Geophys. Res., 77, 567
64. Trussoni, E., in *Numerical MHD and Instabilities*, proceedings of Jetset School and Workshop, Sauze d' Oulx, Italy, 2007
65. Tsinganos, K.C., 1982, ApJ, 252, 775
66. Tsinganos, K., Low, B.C., 1989, ApJ, 342, 1028
67. Tsinganos, K., Trussoni, E., 1991, A&A, 249, 156
68. Tsinganos, K., Sauty, C., Surlantzis, G., Trussoni, E., Contopoulos, J., 1996, MNRAS, 283, 811
69. Tsinganos, K., Bogovalov, S.V., 2000, A&A, 356, 989 (TB00)
70. Tsinganos, K., Bogovalov, S.V., 2002, MNRAS, 337, 553 (TB02)
71. Uchida, Y., Shibata K., 1985, Publ. Astron. Soc. Japan, 37, 515
72. Ustyugova, G.V., et al. 1999, ApJ, 516, 221
73. Vlahakis, N., Tsinganos, K., 1997, MNRAS, 292, 591
74. Vlahakis, N., & Tsinganos, K., 1998, MNRAS, 298, 777
75. Vlahakis, N., Tsinganos, K., Sauty, C., & Trussoni, E., 2000, MNRAS, 318, 417
76. Vlahakis, N., Ko"nigl, A., 2003, ApJ, 596, 1080
77. Washimi, H., 1990, Geophys. Res. Letts., 17(1), 33
78. Washimi, H., Shibata S., 1993, MNRAS, 262, 936
79. Weber, E.J., Davis, L.J., 1967, ApJ, 148, 217
80. Woitas, J., Bacciotti, F., Ray, T.P., Marconi, A., Coffey, D., Eislöffel, J., A&A, 432, 149
81. Zanni, C., Ferrari, A., Massaglia, S., Bodo, G., & Rossi, P., 2004, in *Virtual Astrophysical Jets*, Ap&SS, 293, 99

Transit Flows and Jet Asymptotics

Thibaut Lery

School of Cosmic Physics, Dublin Institute for Advanced Studies, 5 Merrion Square, Dublin 2, Ireland
lery@cp.dias.ie

1 Introduction

The long lasting lack of understanding of star formation is mainly due to the fact that the process has never been directly observed. Indeed, stars begin their lives deep inside dense and cold clouds of molecular gas and dust which hinder optical light from escaping [11]. Fortunately, advances in observational technology over the last quarter century opened the infrared and millimeter-wave windows to astronomical investigation. However, it only enabled observations of indirect phenomena accompanying the star formation process, such as flows and jets, rather than direct observation of the young stellar objects

T. Lery: *Transit Flows and Jet Asymptotics*, Lect. Notes Phys. **723**, 161–180 (2007)
DOI 10.1007/978-3-540-68035-2_7   © Springer-Verlag Berlin Heidelberg 2007

themselves. Hence, is it possible to infer the process of star formation from those flows?

In spite of this difficulty, astronomers were expecting to see the signal of inflowing gas, since, at this stage, the basic energy source is probably gravity. Indeed, matter in the native molecular cloud starts to collapse under the action of gravity and forms a central core that will eventually become a star [2]. Instead, observations have shown that during this accretion phase, part of the gas is rapidly moving away from the central core, at tens of kilometers per second. Usually, two giant lobes of molecules are observed and move in diametrically opposite directions on either side of the protostar [3]. Rather strikingly, their masses are similar to, or even more massive, than that of the protostar itself [4]. Astronomers were, then, facing a paradox: How is a star simultaneously a source of both powerful ejections and infall, and how does a star grow by losing mass? The resolution of such a paradox must be that outflows are driven by infall and accretion [16], as we will see.

Another problem still remains unsolved in the field of star formation. While massive stars are fundamental in the evolution of galaxies since they produce the heavy elements, energize the interstellar medium, and possibly regulate the rate of star formation, little is known about how they form. The problem is difficult observationally because massive star formation occurs in distant, highly obscured regions, and it is difficult theoretically because of the many processes that must be included [4]. Consequently, one can ask whether massive stars are just larger versions of low-mass stars, with the same, but scaled-up, features, and also what determines the fraction of gas in the native cloud that will eventually be transformed into stars.

Thus, the formation of both low and high mass stars, as well as the relation between the infall and the outflows remain open questions, which the present work aims to investigate. To do so, my collaborators and myself have decided to use simplified models of the flows within young stellar objects mainly based on self-similarity [16]. I will start this article by presenting the basis of star formation, of self-similarity, and of our models. I will show the main properties of typical solutions, as well as comparisons with observations. Finally, we will try to investigate potential consequences of the models, such as temporal and spatial evolutions during star formation.

2 Transit Flows

2.1 Coexistence of Infall and Outflow

The birth of a star starts within molecular clouds, by the collapse of the gas that contracts under gravity [11]. The infall is accentuated by the radiation of a fraction of the gas internal energy. At the beginning the gas should contract homogeneously. However, a flow pattern develops that accelerates toward the origin, so precipitating the collapse. At later times, the outer region would have approximatively power-law dependences on the radius. Provided the gas remains optically thin, the wave front goes all the way in. Thereafter a point mass forms and accretes the gas surrounding the origin. At this accretion stage, the central 'protostar' is surrounded by a circumstellar disk and infalling envelope [2]. Well shielded from the ambient interstellar radiation field, the core of the collapsing cloud remains cold (≤ 20 K), detectable in millimetre and sub-millimetre wavelength regime. As the core collapse continues, the core gets heated up by thermal radiation now trapped because of the increased density and optical depth. This slows down the collapse till about 2000 K, the temperature where molecular hydrogen dissociates into atomic hydrogen. This absorbs energy from the gravitational collapse, leading to further collapse of the core. A similar process gets repeated when the temperature reaches to ionize hydrogen atom and later helium. When the radiation pressure grows strong enough to compete against the gravitational pressure, a quasi-stable hydrostatic equilibrium is reached. A pre-main-sequence (PMS) star is born. At this stage such an object can be observed at infrared wavelengths. Meanwhile, the angular momentum increases drastically during the collapse. It is crucial for the angular momentum to be removed in order to further accrete material on the central object. It is speculated that jets and outflows [3], often associated with young stellar objects, can efficiently remove the excess angular momentum. This is a key element to understand the origin of the outflows that surround forming stars.

Observationally it is has been possible to classify the various stages that a protostar has to follow in order to end up as a star [2]. This classification is based on the spectra of low-mass young stellar objects since they are easier to investigate, as previously mentioned. The first stage corresponds to *Class 0* objects, that are the most deeply embedded sources. Such objects are still surrounded by infalling envelopes containing at least half of the mass of the central object. All *Class 0* objects are associated with highly collimated molecular outflows, typically more energetic than those associated with the next stage, i.e. *Class I* objects. The latter are still deeply embedded in dense molecular cores and not optically visible. They are often associated with molecular bipolar outflows, though less energetic than those associated with *Class 0* objects. The *Class II* objects, or Classical T Tauri stars, are surrounded by an accretion disk but with no infalling envelopes. Finally, *Class III* objects have a photo-sphere with a normal stellar wind although free of any significant amounts of circumstellar material (weak-lined T Tauri stars).

Therefore, it is clear that outflows co-exist with infall as protostars form within the collapsing cores of molecular clouds. Infall and outflow both appear to be present from rapidly accreting embedded *Class 0* objects to fully formed T Tauri stars. This suggests that the dynamics leading to the formation of a protostar are more complex than simple radial infall, and are dominated by strongly anisotropic motions.

2.2 Models of Molecular Outflows

The masses of molecular outflows approximately range from 0.1 to 20 solar masses [4]. This indicates that the outflows consist of swept-up ambient material rather than ejecta from a central protostar. One of the fundamental questions here is the nature of the mechanism that sweeps up and accelerates the cold ambient medium into oppositely directed outflows. Here we briefly review current models for the origin of molecular outflows.

The first type of model is the 'wind-driven' model [17, 19]. Such a model postulates that a largely neutral wind is weakly collimated by a thin disk and evacuates a cavity that pushes a shell of the ambient molecular material before it. However, the model fails to explain most of the observed phenomenon.

The second type of model, the 'jet-driven' model, postulates a jet as the prime mover and achieves the collimation and acceleration of the molecular outflow by virtue of entrainment. It seems unlikely as the primary driving mechanism on efficiency grounds. In particular, jet simulations with cooling jets (e.g. [5]) suggest that entrainment becomes much less effective when the jets cool sufficiently that they are 'ballistic' in the ambient medium. A subclass of the previous model consists of outflows driven by intermittent jets either because of sequential ejection or because of 'precession' of the axis. But if the jet were to be the prime mover then it must contain all of the momentum seen in the wider outflow. In some cases this appears plausible, but it is far from clear that this is generally true. Moreover all such models suffer from excessive collimation and from an expectation that the higher luminosity sources should be better collimated.

For the third type of model, the outflow has its origin directly in the disk by virtue of a magnetic field that subtracts gas and angular momentum and deposits it in a collimated outflow (e.g. [6], or [14]). Such 'disk driven' outflow models solve simultaneously in effect the angular momentum and magnetic flux problems of a protostar, as well as allowing the accumulation of its mass, by driving the molecular outflows. There remain some difficulties with this picture however, not the least of which is the amount of energy needed to extract the gas from the disk. Another omission from most of these models to date is the inner boundary condition. However it is virtually certain that the core-disk boundary layer is the seat of energetic activity. It is just this aspect which is emphasized in the next type of model.

In this case, all the gas is ejected in the vicinity of the inner part of the disk. This small scale beamed outflow has presumably to entrain the larger scale molecular outflow. Such boundary layer models are known as 'X-wind' models [18]. There seem to be a number of outstanding problems however. Perhaps the most difficult is the cross field line flow that is required in their innermost region. Moreover some unspecified torque coupling is assumed to act across this region, which begs some of the important questions. The model has not yet been carried to the point of a direct comparison with radio observations.

A last and recent type of model, that is the subject of the present work, takes into account the heating of the central object for the deflection and acceleration of the molecular gas [7, 9, 13]. In this heated quadrupolar self-similar model, as material falls towards the central object, it is gradually slowed down by increasing temperature, density and magnetic field encountered by material near the central object. The pressure barrier, with the help of centrifugal forces, deflects and accelerates much of the infalling matter into an outflow along the axis of symmetry. The molecular material could eventually be entrained by the central fast outflow, described in all the previous models. In this case, a smaller momentum would be needed for the entrainment.

2.3 Origin of the Quadrupolar Geometry

Most of current outflow models are based on a dipolar geometry where the magnetic field lines are either advected via accretion towards the central object in the accretion disk or are created in situ. Here we suggest a plausible scenario for the origin of the magnetic structure of our model, where both dipolar and quadrupolar magnetic field can be generated during the earliest stages of the formation of the protostar. Figure 1 shows such a temporal sequence taking place in the parent molecular cloud. At the first stage, the magnetic field is mainly poloidal and the gas starts to collapse. The infalling material concentrates the magnetic field towards the accreting core mostly in the equatorial region. This process should occur quasi-isothermally until the optical depth in molecular lines approaches unity, carbon-monoxide being the main coolant in the cloud. As the central object accretes mass and starts heating, the pressure 'barrier' described above starts affecting the dynamics and deflects some of the gas. The magnetic field is advected with the deflected gas, forming high magnetic arches as shown in the third panel of Fig. 1. The accretion disk appears as well as the jet. Of course, the medium has to be resistive in order for reconnection to take place in the turbulent and heated gas and finally for the quadrupolar structure to settle. At last, the inner accretion disk and the jet still remain with a dipolar structure while the circulation region ends up with a quadrupolar structure.

In addition to this potential scenario, any strong differential rotation of accretion disks can be responsible for the excitation of quadrupolar modes in disks, since accretion disks near protostars are probably heavily convective and therefore prone to dynamo action. This is particularly true for rapidly rotating disks, where this dynamical system may evolve naturally into a quadrupolar structure [10]. Note that even though the poloidal components of the magnetic and velocity fields have to be parallel, the toroidal components need not share the same constant of proportionality. This permits a poloidal conservative electric field to exist in the inertial frame, and so admits steady Poynting flux driving. One needs then to introduce an electric potential function and an azimuthal magnetic field independent of the azimuthal velocity.

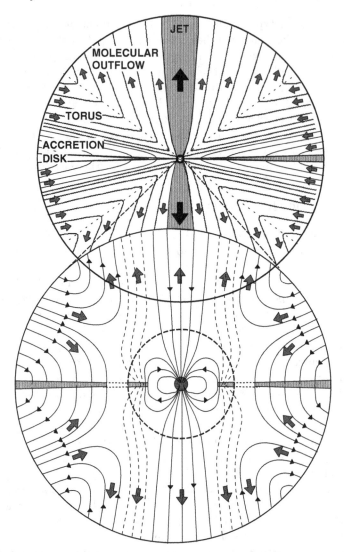

Fig. 1. **Schematic evolution of a collapsing molecular cloud towards quadrupolar topology.** Heavy solid lines correspond to the magnetic field surfaces. The grey arrows represent the motion of the gas. The central grey circular region (about 100 AU), that represents the zone where the jet is launched, is not treated in the present model as well as the thin accretion disk (*also in grey*). The protostar is located at the centre of this circle

These considerations have led us to use the set of equations of steady, axisymmetric, ideal MHD and to seek solutions with a quadrupolar geometry, in order to model outflows around protostars [16]. The last but crucial element of the model is the self-similarity, that we are going to present now.

2.4 Self-similarity

A phenomenon is called self-similar if the spatial (or temporal) distributions of its properties at various different times (or locations) can be obtained from one another by a similarity transformation. This means that the investigation of the full phenomenon can be reduced to the study of the properties of the system for only a specific time (or location). Thus, for example, if the density of the distribution of matter is known everywhere in space at a given time, then it is known at any subsequent time. If the origin of time can be chosen arbitrarily, the scales of length and mass are also arbitrary, and the system is 'scale-free'. This simplifies the problem drastically. Mathematically, it implies a reduction of the system of partial differential equations that describe the system, to ordinary differential equations, which most of the time makes their investigation simpler.

For a long time, self-similar solutions were treated by most researchers as though they were merely isolated exact solutions to special problems: elegant, sometimes useful, but extremely limited in significance. In fact, it was only gradually realized that these solutions were actually of much broader significance. They turn out not only to describe the behavior of physical systems under some special conditions, but also describe the intermediate-asymptotic behavior of solutions to wider classes of problems in the range where these solutions no longer depend on the details of the initial and/or boundary conditions, yet the system is still far from being in an ultimate equilibrium state. Self-similar solutions have also served as standards in evaluating approximate methods for solving more complicated problems. Such self-similar solutions provide, at least, the basic physical insights for the general problems, and may indicate the way to more detailed investigations.

The crucial step in any field of research is to establish what is the minimum amount of information that is actually required about the phenomenon being studied. All else should be put aside in the study. Thus, the primary thing in which the investigator is interested is the development of the phenomenon for intermediate times and distances away from the boundaries such that the effects of random initial features or fine details in the spatial structure of the boundaries have disappeared but the system is still far from its initial equilibrium state. This is precisely where the underlying laws governing the phenomenon appear clearly. Therefore intermediate asymptotics are of primary interest in every scientific study. It is noteworthy that the concept of intermediate asymptotics is used not only in mathematical physics, but also in our perception of visual art, for instance. We have to look at paintings at distance great enough not to see the brush-strokes, but at the same time, small enough to enjoy not only the paintings as a whole but also its important details.

There exist two kinds of self-similar models. First, the self-similar solutions of the first kind are obtained when in the passage to the limit from the non-self-similar non-idealized problem to the self-similar idealized problem, there is a complete similarity in the parameters that make the problem non-idealized and its solution non-self-similar. Expressions for all the self-similar variables can be obtained by applying dimensional analysis. For the solutions of the second kind, the idealization of the original problem is such that there is incomplete similarity in the similarity parameters. Determination of the exponents of the self-similar variables leads to a non-linear eigenvalue problem. The constant multiplier appearing in the self-similar variables is left undetermined in the direct construction of the self-similar solutions of the second kind. This constant can be found by flowing the entire process of evolution of a solution of the non-idealized problem into a self-similar asymptotics.

In the present problem of star formation, considerations of dimensional analysis turn out to be sufficient for proving the self-similarity of the second kind starting from the formulation of the mathematical problem, and for obtaining expressions for the scale and the self-similar variables. In our case, self-similarity means that we can choose variable scales such that in the new scales the properties of the phenomenon can be expressed by functions of one variable, for example $F(r, \theta) = f_1(\theta) f_2(r)$. The function that determines the factor by which the numerical value of a physical quantity changes upon passage from the original system of units to another system within a given class is called the dimension function of that quantity. The dimension function for any physical quantity is always a power-law monomial. This follows from a single, naturally formulated (but actually very deep) principle: all systems within a given class are equivalent, i.e. there are no distinguished, somehow preferred, system among them. Therefore, if we assume that $f_2(r)$ is given by a power law of r, the solutions of the problem can be reduced to the solution of a system of ODEs for the vector function $f_1(\theta)$.

Furthermore, it is possible to separate dimensional quantities from the dimensionless ones. Indeed, in our case, we can consider that the central object, that will eventually become a star, releases thermal energy not at a point, but within a sphere of a certain finite radius r_o. If we choose such a scale r_o for the radial variable and F for any property of the phenomenon, then the distribution of $\mathbf{F}(r, \theta)$ can now be expressed in the form

$$\mathbf{F}(r, \theta) = \mathbf{F_o} \, f_1(\theta) \, f_2(r/r_o) \, . \tag{1}$$

In this equation, $\mathbf{F}(r, \theta)$ and $\mathbf{F_o}$ are dimensional quantities. Such a self-similar solution to the problem with singular initial data is an asymptotics of a wider class of solutions of initial-value problems, up to a constant, $\mathbf{F_o}$ in the present case.

Let us see now in more detail how we can apply this formalism to the infalls and outflows surrounding young stellar objects.

2.5 Transit Models

A very interesting indication on the application of self-similarity has come from numerical simulations a few years ago. Indeed, Tomisaka [20] has studied numerically the dynamical collapse of magnetized molecular cloud cores from the runaway cloud collapse phase to the central point mass accretion phase. He has found that the evolution of the cloud contracting under its self-gravity is well expressed by a self-similar solution. Moreover inflow-outflow circulation appeared as a natural consequence of the initial configuration. Such a result suggests that the self-similar approach can be a good first approximation of the infalls and outflows around protostars. Indeed, the self-similar heated, quadrupolar and axisymmetric magnetohydrodynamic models regard the molecular outflow as a natural consequence of the circulation established by the collapse of the pre-stellar cloud. Let us now describe this model in more detail.

The Variables

The solutions are developed within the context of r-self-similarity wherein a power of r multiplies an unknown function of θ; the spherical coordinates r, θ and ϕ being used [16]. The only physical scales that enter into our calculation are the gravitational constant G, the fixed central mass M, and a fiducial radius r_o [13]. The power laws of the self-similar system are determined, up to a single parameter α, if we assume that the local gravitational field is dominated by a fixed central mass. In terms of the fiducial radial distance, r_o, the self-similar symmetry is sought as a function of two scale invariants, r/r_o and θ, in a separated power-law form. The self-similar index α is a free parameter of the solution, but must lie in the range $-1/2 < \alpha \leq 1/4$, for simultaneous infall and outflow to occur [16].

Hence, if we assume that the gravitational potential is dominated by the central mass, i.e. self-gravitation is negligible, the equations of radiative MHD admit the following radial scaling relations for the variables

$$\mathbf{v} = \sqrt{\frac{GM}{r_o}} \left(\frac{r}{r_o}\right)^{-1/2} \mathbf{u}(\theta) , \tag{2}$$

$$\rho = \frac{M}{r_o^3} \left(\frac{r}{r_o}\right)^{2\alpha-1/2} \mu(\theta) , \tag{3}$$

$$B_{\phi,p} = \sqrt{\frac{GM^2}{r_o^4}} \left(\frac{r}{r_o}\right)^{\alpha-3/4} \frac{u_{\phi,p}(\theta)}{y_{\phi,p}(\theta)} , \tag{4}$$

$$p = \frac{GM^2}{r_o^4} \left(\frac{r}{r_o}\right)^{2\alpha-3/2} P(\theta) , \tag{5}$$

$$\frac{kT}{m_\mu m_H} = \frac{GM}{r_o} \left(\frac{r}{r_o}\right)^{-1} \Theta(\theta) , \tag{6}$$

$$\mathbf{F}_{rad} = \left(\frac{GM}{r_o}\right)^{3/2} \frac{M}{r_o^3} \left(\frac{r}{r_o}\right)^{\alpha_f-2} \mathbf{f}(\theta) , \tag{7}$$

where $\mathbf{v}$, ρ, $\mathbf{B}$, p, T and $\mathbf{F}_{rad}$ respectively correspond to the velocity, the density, the magnetic field, the pressure, the temperature and the radiative flux. In these equations the microscopic constants are represented by k for Boltzmann's constant, m_μ for the mean atomic weight, m_H for the mass of the hydrogen atom. In the last equation, the index α_f is a measure of the loss (if negative) or gain (if positive) in energy by radiation as a function of radial distance. Indeed, when α_f is negative, it corresponds to a loss in energy due to the net radiation leaving the region, while, for a positive α_f, the system has an energy input due to the absorption of radiation. If one supposes that the opacity is predominantly due to dust, the index α_f is related to α by $\alpha_f \approx -2(1/4 - \alpha)$. Consequently, the case $\alpha_f = 0$, where there is no net flux of radiation through the domain of our solution, corresponds to $\alpha = -1/4$. For larger (respectively smaller) values of α, there is a net loss (gain) of energy by radiation [7, 13, 16].

The self-similar variable directly related to magnetic field y can be divided into poloidal and toroidal components (see Fig. 2). In the present model the two components are not equal. They are respectively defined by $y_{p,\phi}(\theta) = M_{ap,a\phi}/\sqrt{4\pi\mu(\theta)}$. Consequently the system also deals with two different components of the Alfvénic Mach number M_{ap} and $M_{a\phi}$ defined by $M^2_{ap,a\phi}(\theta) \equiv v^2_{p,\phi}/\left(B^2_{p,\phi}/4\pi\rho\right)$. We will refer to Θ_0 as the value of the self-similar temperature $\Theta(\theta)$ on the axis.

The System of Equations

We use the self-similar forms in the usual set of ideal MHD equations together with the radiative diffusion equation when applicable [16]. In order to make the system tractable, we assume axisymmetric flow so that $\partial/\partial\phi = 0$ and all flow variables are functions only of r and θ. We further restrict ourselves to steady models (i.e. $\partial/\partial t = 0$). Magnetic field and streamlines are required to be quadrupolar in the poloidal plane for the circulation model. Under these assumptions the self-similar equations are:

1. Mass flux conservation

$$(1 + 2\alpha)\, \mu\, u_r + \frac{1}{\sin\theta} \frac{d}{d\theta} (\mu\, u_\theta \sin\theta) = 0 \,, \tag{8}$$

2. Magnetic flux conservation

$$\frac{(\alpha + 5/4)\, u_r}{y_p} + \frac{1}{\sin\theta} \frac{d}{d\theta} \left(\frac{u_\theta \sin\theta}{y_p} \right) = 0 \,, \tag{9}$$

3. Radial component of momentum equation

$$u_\theta \frac{du_r}{d\theta} \left(1 - M_{ap}^{-2}\right) - \left(u_\theta^2 + u_\phi^2\right)\left(1 - \frac{\alpha + 1/4}{M_{ap}^2}\right)$$
$$- \frac{u_r^2}{2} - \left(\frac{3}{2} - 2\alpha\right)\Theta + 1 + \frac{u_r u_\theta}{M_{ap}^2 y_p} \frac{dy_p}{d\theta} = 0 \,, \tag{10}$$

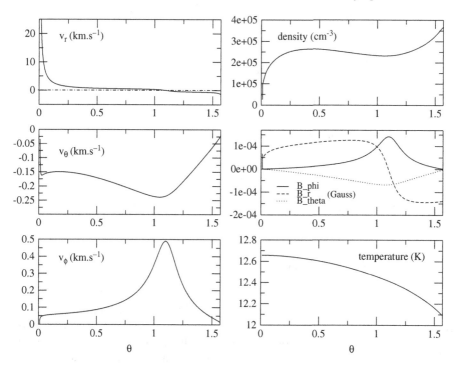

Fig. 2. Poloidal current circuit in the asymptotic domain. A schematic representation of the current, in the case of a field with two different polarity zones (*dipolar symmetry, left*) on the wind source (central object), and in the case of three different polarity zones (*quadrupolar symmetry, right*) on the wind source. Current is concentrated in boundary layers near the pole and the neutral magnetic surfaces (*dotted lines*). The heavy arrows indicate the main poloidal electric current channels. The dashed line is meant to represent a surface at infinity. In the case of kinetic winds there is also a very weak and diffuse current in the field regions between the regions of largest current flow so that the boundary layer current vanishes at infinity

4. θ-component of momentum equation

$$\frac{u_r u_\theta}{2}\left(1 - \frac{2\,\alpha + 1/2}{M_{ap}^2}\right) + u_\phi^2 \cot\,\theta\left(M_{a\phi}^{-2} - 1\right)$$

$$+\frac{u_r}{M_{ap}^2}\frac{du_r}{d\theta} + u_\theta\frac{du_\theta}{d\theta} + \frac{u_\phi}{M_{ap}^2}\frac{du_\phi}{d\theta} - \frac{u_r^2}{y_p M_{ap}^2}\frac{dy_p}{d\theta}$$

$$-\frac{u_\phi^2}{y_\phi M_{a\phi}^2}\frac{dy_\phi}{d\theta} + \frac{d\Theta}{d\theta} + \frac{\Theta}{\mu}\frac{d\mu}{d\theta} = 0\,, \qquad (11)$$

5. Angular momentum conservation

$$\frac{1}{u_\phi}\frac{du_\phi}{d\theta}\left(1-\frac{1}{M_{ap}M_{a\phi}}\right)+\frac{u_r}{u_\theta}\left(1/2-\frac{\alpha+1/4}{M_{ap}M_{a\phi}}\right)$$

$$+\cot\theta\left(1-\frac{1}{M_{ap}M_{a\phi}}\right)+\frac{1}{y_\phi M_{ap}M_{a\phi}}\frac{dy_\phi}{d\theta}=0\,,\tag{12}$$

6. Faraday's Law plus zero comoving electric field

$$\frac{d(u_\phi u_\theta)}{d\theta}+\left[\alpha-\frac{1}{4}\right]u_\phi u_r+u_\phi u_\theta\frac{d}{d\theta}\ln\left[\frac{1}{y_p}-\frac{1}{y_\phi}\right]=0\,.\tag{13}$$

This set of equations can produce either circulating flows, pure infall, or pure outflow.

Typical Behavior of Solutions

Here, we list the main features that solutions of the models show. The principal characteristic of the model is that it produces a heated pressure-driven outflow with magneto-centrifugal acceleration and collimation. An evacuated region exists near the axis of rotation where the high speed outflow is produced [7]. This outflow decreases in speed and increases in mass systematically with angle from the axis. Near the equatorial plane a thick rotating extended disk forms naturally when sufficient heating is provided to produce a high-speed axial outflow. The most rapidly outflowing gas is always near the symmetry axis because these streamlines pass closest to the star, deeper into the gravitational potential well. Also, the material on these streamlines is heated the most vigorously. As the gas gets closer to the source, it rotates faster. Gas streamlines make a spiraling approach to the axis and then emerge in the form of an helix wrapped about the axis of symmetry. The infalling plasma therefore has a larger electric current driven by the rotational motion. This increases the magnetic energy at the expense of gravity and rotation, which is eventually converted into kinetic energy as the gas is redirected outwards [16]. The magnetic field acts to collimate and accelerate the gas towards the polar regions. There the flow presents a strong poloidal velocity and a low magnetic energy. The Poynting flux included in the model increases both the velocity and collimation of the outflows by helping to transport mass and energy from the equatorial to the axial regions.

One remarkable prediction of these models is that the magnetic field at a given radius varies dramatically with angle. The values range from 10 micro-Gauss to a milliGauss at 10^4 AU, from 10^{-2} to a few Gauss at 20 to 40 AU in agreement with observations [3]. Peak field strengths may reach values as high as 1 to 100 Gauss at 1 AU.

It is found that massive protostars produce faster velocities with large opening angles. Rotation and collimation decrease together with increasing temperature and therefore probably with the bolometric luminosity of the central object. It is also found that larger opening angles are associated with smaller magnetic fields. If the magnetic field varies secularly as the evolution proceeds, then a sequence of our models could be regarded as a series of 'snapshots' of the protostellar evolution. This evolutionary sequence would show that the opening angle increases with time, as the magnetic field becomes less important. This would be consistent with the notion that the youngest outflows are generally the most collimated.

An important point is that the material ejected in the outflow acquires an important energy due to a combination of heating by the central protostar, pressure gradients developed by the infalling material, and magnetic forces due to the non-zero Poynting flux.

Comparison with Observations

Synthetic spectral lines from ^{13}CO (J=1 → 0) allow direct comparison with observational results via channel maps, maps of total emission, position-velocity and intensity-velocity diagrams [7, 13]. Synthetic radio maps of total intensity and in velocity channels have been calculated based on the dynamical solution, projected on the sky at various angles. Most of the features of outflow observations are reproduced in almost every respect. Not only the forms of the lines but also their brightness temperatures are reproduced and this in turn leads to predictions for densities, dust temperatures and magnetic field strengths. The most important feature of these channel maps is that the opening angle of the outflow gradually decreases as the magnitude of the velocity increases. Thus, our models naturally produce outflows in which most of the material is poorly collimated and moves at relatively low velocities, but the fastest jet-like components are very well collimated towards the axis of symmetry.

Many molecular outflows show wide, hollow cavities at the base of the outflow. In the present model, the molecular cavities may be identified with the substantial decrease in density in the intermediate region between the jet and the molecular outflow. The central jet is in atomic form, being of much higher excitation, and occupies the axial region. The molecular outflows appear thus as a hollow conical structure. In this scenario, the cavity is a result of the circulation pattern itself, and we need not assume that the jet is precessing. Of course, the action of the jet may also participate the widening of the cavity with time.

The circulation model provides a self-sustained acceleration of the molecular material in the axial region. This has interesting consequences concerning the subsequent interaction of the flow with the coaxial jet. Since in our case, the difference in velocities between the jet and the molecular outflow material are reduced from the start, the shocks in the zone of acceleration due to their interaction should be less strong. This would also imply that the post-shock cooling time is reduced too. In this way, the kinetic temperature in the outflow would rapidly decrease again to a value comparable to that of the ambient medium as shown from observations.

The central fast jet has still the largest part of the total momentum per unit area, and the molecular outflow could undergo a prompt entrainment from the head of the jet. But the most interesting feature of the circulation model is probably that it can produce solutions where the mass of the molecular outflow is larger than the final mass of the forming star. This is particularly true if self-gravity is included in the model. One may then understand how bipolar outflows from massive protostars are observed to transport masses largely exceeding those of the associated stars.

3 Jet Asymptotics

3.1 General Results

The last but central element around protostars is the fast jet. A consensus seems to prevail on the magneto-centrifugal origin of jets, either launched from the accretion disk (*disk wind*) or from the location of the interaction of the protostar's magnetosphere with the disk (*X-wind*). Indeed, as we have mentioned earlier, accretion disks play a key role in the physics of the fast jets from young stellar objects. Infalling, rotating matter is stored in these disks until dissipation allows material to spiral inward and feed the central, gravitating object. Such disks are believed to support strong, well ordered magnetic fields. The current consensus holds that these fields are the agents for producing jets in a process known as magneto-centrifugal launching. In this mechanism, plasma in the disk is loaded on to corotating field lines. If conditions in the disk are favorable the plasma is centrifugally flung outward along open field lines, which form a certain angle with the disk's surface. The ensuing plasma flow properties must then be determined by solving for the equilibrium of forces parallel and perpendicular to the magnetic surfaces, the former described by using the Bernoulli equation and the latter is solved via the Grad-Shafranov equation.

We have proposed a model for jet, known as the Given Geometry Method [12, 14] that allows asymptotic MHD jet equilibria to be linked directly to the properties of the rotating source. The model assumes a time-independent, axisymmetric flow with a polytropic equation of state. It further simplifies the problem of magneto-centrifugal launching/collimation by assuming that the nested magnetic flux surfaces defining the flow possess a shape which is known a priori inside the fast critical surface. The fast surface defines the locus of points beyond which the flow is kinetic energy dominated. The flux surfaces are assumed to be conical and, as an additional simplification, an equilibrium across the surfaces is assumed at the Alfvén point which yields an equation referred to as the Alfvén regularity condition. The equilibrium parallel to the surfaces takes the form of criticality conditions at the two other critical points. Boundary conditions are provided as a model for the source rotator. Asymptotically, the jet is assumed to be in pressure equilibrium with the external medium. The pressure matching condition along with with the Grad-Shafranov and Bernoulli equations are all solved in the asymptotic cylindrically collimated regime.

The solutions have the following general properties:

- For winds which are kinetic-energy dominated at infinity the analytic solutions have magnetic surfaces that focus into paraboloids. The current slowly weakens as the inverse of the logarithm of the distance to the wind source while the axial plasma density falls-off as a negative power of this logarithm.
- For winds carrying Poynting flux at large distances the solutions asymptotically approach to nested cylindrical and conical magnetic surfaces.

All winds have been shown to possess a circumpolar current-carrying boundary layer, which has the structure of a pressure-supported plasma-jet pinch. Null-surface boundary layers have the structure of pressure-supported current sheets. The total electric current is constant or slowly diminishes with distance according to an inverse logarithmic law for the Poynting flux and kinetic winds respectively. This diminution is caused by minute amounts of current flowing through the diffuse field regions from the pole to the nearest null surface.

3.2 Jet Simulations

Establishing initial equilibria for MHD jet simulations is non-trivial. This is not the case for hydrodynamic jets where the required force balance across the jet and ambient medium interface allows for the use of so-called *top-hat* profiles (i.e. the hydro variables are constant across the jet cross section). Such distributions may not be tenable in MHD jet studies. The difficulty can be seen by decomposing the Lorentz force into a tension term and a pressure term. $\mathbf{F}_l \propto -\nabla B^2 + 2(\mathbf{B} \cdot \nabla)\mathbf{B}$. In a steady, cylindrically collimated jet only B_ϕ and B_z components of the field are possible. Jets with purely longitudinal fields, $B = B_z$, can be easily set in pressure balance with the environment and top-hat profiles may be used. Toroidal or helical field geometries require more complicated initial conditions unless the field is assumed to take on a force free configuration. If the field is not force free, MHD jets must have variable distributions of gas pressure and, perhaps, other variables in order to balance the *hoop stresses* associated with the tension force.

Faced with the problem of initial conditions researchers studying radiative MHD jets have, in general adopted one of two strategies: (i) use force-free fields; (ii) use ad-hoc gas pressure and magnetic field distributions configured to be in initial force balance. The results of these studies for both steady and time-variable (pulsing) jets reveal a number generic features.

- Jets with purely longitudinal geometries do not show propagation characteristics which differ significantly from the hydrodynamic case.
- Jets with a toroidal field component will be subject to strong hoop stresses especially in the region between the jet and bow shocks.

There is a third approach that we have used to simulate jets [14] whose initial conditions jets are taken directly from the model of the magneto-centrifugal launching/collimation process described above. The model, known as the Given Geometry Method (GGM) allows asymptotic MHD jet equilibria to be linked to the properties of a rotating source.

In terms of dynamics, The most important properties of the jets derived via the GGM are the radial variations in the density $\rho(r)$ and the toroidal component of the magnetic field $B_\phi(r)$. In particular when the source is composed of a rapidly rotating disk truncated some distance from a rigidly rotating star, the emitted jets can have strong density stratifications, i.e. a high density axial *core* surrounded by a lower density *collar*. The strongest toroidal field is located at the boundary of the core and collar creating a magnetically confined jet-within-a-jet structure. Note that the bulk of the jet's momentum resides in the core. Hence we expect this portion of the beam to penetrate more easily into the ambient medium during the jet's propagation while the collar will be more strongly decelerated.

The jets in these simulations are considerably more complex than the usually used 'top-hat' profiles [8]. Many features of the simulation are in good agreement with observations, such as the molecular cavities, the location and shape of the shocks, as well as the variation with distance of the ionization fraction and of the density along the jet. The stability analysis of the equilibria used in the simulations has also been carried out [15], allowing us to investigate the magnetic current driven instabilities that develop in the jet simulations. By varying the properties of the source, it is also possible to vary the properties of the jet itself. This introduce non-ad-hoc variations of the jet and gives rise to more complex behaviors of the propagating jet but also of the interaction with the ambient medium. This opens the possibility that the physics of the jet source may be read off the jets themselves. Our results suggest that one might ideally be able to distinguish between different classes of MHD launching models via consideration of the way the jets from these models would appear on the sky.

4 Global Infall-outflow Models

Star forming regions show many different types of flows and elements, optical jets, molecular outflows, infalls, accretion disk, a central protostar, masers and other emissions. These are clearly inter-related, but is it because they have the same cause, or because one or more components provoke the others, or because they are simply different aspects of the same underlying global behavior?

With this question in mind, we have proposed a global model [16] that combines a jet model, a circulation model and an infalling envelope model. In the vicinity of the protostar, we have a accretion-ejection engine driving the jet. The molecular outflow is primarily powered by the infalling matter. The outflow follows a circulation pattern around the central object without being necessarily entrained by the jet as schematically represented in Fig. 3. The last element of the model is the extended region where material is falling from the cloud onto the flattened accretion disk, not treated by the present model. This falling envelope extends to a few thousand AU and is also described by the set of self-similar equations of the circulation model. In fact, the material, for both the circulation and the infalling envelope models, originate from the same region in the molecular cloud. We have suggested that the same set of equations can apply in both cases. This set of equations would not be applicable to the thin disk where resistivity, ionization fraction and other parameters can substantially differ from the outer regions. In fact, this thin disk acts as a sink for the infalling gas in our model.

The only physical scales that enter into our calculation are the gravitational constant G, the fixed central mass M, and a fiducial radius r_o. The

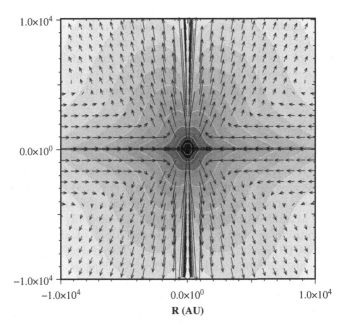

Fig. 3. Streamlines around protostars. Schematic representation of streamlines and magnetic field lines projected in the poloidal plane. Large and small arrows respectively indicate the matter trajectories and the magnetic field directions. The typical sizes of the upper and lower views, and of the dashed circle, enlarged for clarity, are 2,000, 200 and 0.5 AU respectively. The zones respectively correspond to the jet, the circulation and the infalling envelope models

parameters of the model are the indexes α and α_f of the self-similar system, and θ_{min} and θ_{max}, the limiting angles between the three regions of the model. Magnetic field and streamlines are required to be quadrupolar in the poloidal plane for the circulation model.

5 Time Dependent Models

We have also investigated time dependent and anisotropic collapse models for earlier stages of stellar formation when the star has not yet formed, and most of the gas still resides in the surroundings [1]. These self-similar models take quite a different approach by treating the time-dependent problem of accretion and simultaneous outflow in a dynamically collapsing and self-gravitating core. The basic non-dimensional quantities from which we construct our self-similar model are given by the poloidal angle θ and the variable $X \equiv -\frac{r}{c_s\,t}$, where r is spherical radius, t is time, and c_s is a *fiducial* sound speed. The models provide a reasonably complete description of the dynamics on all scales between the inner hydrostatic core and an outer X point. The previous steady-state version of the model, described in the previous sections, is expected to apply external to the collapsing region modeled here, and possibly at later times. Remarkably, such a collapse model that includes self-gravity, time-dependence, rotation and magnetic field, admits an exact and completely analytic solution. We note that there are few other analytic solutions of this complexity in all of MHD. The main point of this work was to demonstrate that infall and outflow can coexist and arise naturally from our self-similar equations, especially during the earliest stages of formation of the central protostellar core.

6 Conclusion

In the introduction, we were asking how a star could simultaneously be a source of both powerful ejections and infall, and how a star could grow by losing mass? We have seen that such a paradox is solved if the outflows are driven by infall and accretion. In this respect, we have presented a global MHD model for flows around young stellar objects. The model is based on the self-similarity assumption applied to the basic equations of ideal axisymmetric and stationary MHD, including Poynting flux. The global model combines a jet model, a circulation model and an infalling envelope model. Instead of the usual mechanisms invoked for the origin of molecular outflows, the outflow is powered by the infalling matter through a heated quadrupolar circulation pattern around the central object. The solutions show dynamically significant density gradients in the axial region, precisely where the radial velocity and collimation are the largest. From an observational point of view, we have clarified the nature of the molecular outflow acceleration and its relation with the fast jet, by providing a global picture of the jet/outflow system which does not primarily rely on entrainment (prompt or turbulent). The model also reproduces well observational features.

In the introduction, we were also asking whether massive stars were just larger versions of low-mass stars. The answer that we have presented here is not simple. Indeed, the various elements in the vicinity of the protostar, i.e. accretion disk, jet, molecular outflow, infalling envelope, are present in both cases, but they have different importance. The central jet engine appears to be dynamically more important for low mass stars, while it is the circulation part that gives rise to the molecular outflows that dominates the properties of massive objects.

In conclusion, we suggest that molecular outflows are dominated by the global circulation of material around the protostar, except for in a thin layer surrounding the jet, where the dynamics is governed by entrainment. We stress that, in the present model, the two flows (atomic and molecular) are not strongly linked dynamically, and, hence, there is no need to transfer large momentum from the jet to the molecular outflow through the entrainment processes.

The present work suggests that radiative heating and magnetic field may ultimately be the main energy sources driving outflows during star formation, at the expense of gravity and rotation. Finally, although the details of the outflows mechanism may be peculiar to individual objects we believe the infall-outflow circulation to arise naturally given accretion, and thus could also be present in other astronomical objects such as active galaxies and around suitably placed compact objects, such as neutron stars.

7 Open Questions

We would like to thank the organizers of the school for their work and the Cosmogrid Project funded by the Irish Higher Education Authority. To conclude, we now list open questions that will hopefully be addressed in a near future by students and young researchers. How does a star set its mass? Do all stars produce jets? Do jet variations come from the source or from instabilities? Could observations of jets allow us to discriminate between models? Can we reproduce jet physics in laboratory experiments?

References

1. Aburihan M., Fiege J.D., Henriksen R.N., Lery T. 2001, MNRAS, 326, 1217.
2. André P., Ward-Thompson D., Barsony M. 1993, ApJ, 406, 122.
3. Bachiller R. 1996, ARA&A, 34, 111.
4. Churchwell E. 1997, ApJ, 479, 59.
5. Downes T.P., Ray T.P. 1999, A&A, 345, 977.
6. Ferreira, J. 1997, A&A, 319, 340.
7. Fiege, J.D. and Henriksen, R.N. 1996, MNRAS, 281, 1038.
8. Frank A., Lery T., Gardiner T.A., Jones T.W., Ryu D. 2000, ApJ, 540, 342.

9. Henriksen R.N. and Valls-Gabaud, D. 1994, MNRAS, 266, 681.
10. Khanna, R., & Camenzind, M. 1996, A&A, 307, 665.
11. Lada C.J. 1991, in The Physics of Star Formation and Early Evolution, eds. CJ Lada, ND Kylafis. Kluwer: Dordrecht.
12. Lery T., Heyvaerts J., Appl S., Norman C. 1999, A&A, 347, 1055.
13. Lery T., Henriksen R.N., Fiege J.D. 1999, A&A, 350, 254.
14. Lery T., Frank, A. 2000, ApJ, 533, 897.
15. Lery T., Baty H., Appl S. 2000, A&A, 355, 1201.
16. Lery T., Henriksen R., Fiege J., Ray T., Frank A., Bacciotti F. 2002, A&A, 387, 187.
17. Shu F., Ruden S., Lada C.J., and Lizano S. 1991, ApJ, 370, L31.
18. Shu F., Najita J., Ostriker E., Wilkin F., Ruden S., & Lizano S. 1994, ApJ, 429, 781.
19. Snell R.N. 1985, Can J Phys, 64, 431.
20. Tomisaka, K. 1998, ApJ, 502, 163.

MHD Disc Winds

Jonathan Ferreira

Laboratoire d'Astrophysique de Grenoble, BP 53, 38041 Grenoble Cedex, France
Université Joseph Fourier, Grenoble, France
Jonathan.Ferreira@obs.ujf-grenoble.fr

1 The Accretion-ejection Paradigm

This lecture is designed to be read with an accompanying file (pdf or ppt) where more illustrations and figures can be found. It can be retrieved at the URL: *http://www-laog.obs.ujf-grenoble.fr/~ferreira/JETSET/school.html*. I also recommend the reviews of Königl & Pudritz [32], Ferreira [17]. In Ferreira et al. [19], a review of all MHD models for Young Stellar Objects has been made with a comparison of the corresponding jet kinetic observational properties.

1.1 A "Universal" Picture

Actively accreting "classical" T Tauri Stars (TTS) often display supersonic collimated jets on scales of a few 10–100 AU in low excitation optical forbidden lines. Molecular outflows observed in younger Class 0 and I sources may be powered by an inner unobserved "optical jet" (see Cabrit's contribution, this volume). These jet signatures are correlated with the infrared excess and accretion rate of the circumstellar disc [6, 27]. It is therefore widely believed that the accretion process is essential to the observed jets, although the precise physical connection remains a matter of debate: do the jets emanate from the star, the circumstellar disc or the magnetospheric star-disc interaction?

One argument in favor of accretion-powered disc winds is its "universality" [34]. Indeed, self-collimated jet production from accretion discs is also invoked to explain an accretion-ejection correlation observed in compact objects (i.e. some active galactic nuclei, quasars and X-ray binaries, see e.g. [39] and references therein). The underlying idea is quite simple: accretion discs around a central object can, under certain circumstances and whatever the nature of this object, drive jets through the action of large scale magnetic fields. These fields would tap the mechanical energy released by the mass accreting in the disc and transfer it to the fraction that is ejected [5]. The smaller the fraction and the larger the final jet velocity. One thing that must be understood is how the presence of such jets modifies the nature of the underlying accretion flow. Many papers in the literature actually assume (implicitly or not) that the accretion disc resembles a standard accretion disc as envisioned by Shakura & Sunyaev [49]; Frank et al. [24]. This is wrong as will be shown later.

A Magnetized Accretion-Ejection Structure (hereafter MAES) is an accretion disc where accretion and ejection are interdependent processes. As such, it is composed of an accretion disc (called hereafter JED for Jet Emitting Disc) thread by a large scale magnetic field of bipolar topology and giving rise to the two bipolar jets. The goal of the study of a MAES is to obtain

(1)- the conditions allowing for a steady state accretion-ejection process;

(2)- the ejection to accretion rates ratio as function of the disc physical conditions;

(3)- the jet properties (kinematics, power, shape) as function of the disc properties.

1.2 From Magnetostatics to Magnetohydrodynamics

Magnetohydrodynamics (MHD) is the theoretical framework required to describe the interaction between an ionized gas and magnetic fields. But magnetostatics is very helpful to understand basic mechanisms.

A zeroth order description of a MAES is that of a rotating conducting disc thread by a magnetic field aligned with the rotation axis (much alike a Barlow's wheel). According to Faraday's induction law, an electromotive force (emf) across the disc, $e = \int (u \wedge B) \cdot dr = \int \Omega r B_z dr$, creates an electric potential difference between the disc center and its border (Fig. 1). If some conducting wire connects the border to the center, closing thereby an electric circuit, then a radial electric current is induced. Because of this current I, the disc becomes prone to a Laplace force, $F = \int I B_z dr$, which will slow down the disc (Lenz's law). One could also say that the field "resists" to the shear provoked by the rotation (the current I induces a toroidal component B_ϕ). But such a "mechanical" view of the magnetic field disregards its electromagnetic nature and one may tend to forget that electric currents must be maintained and able to flow.

In astrophysics, the disc is made of gas that, provided it can cross the field lines, will accrete towards the central object as it looses angular momentum. This angular momentum is linked to the electric current flowing in the jets: the jet kinetic power is fed by the flux of magnetic energy provided at the disc surface. Note that while the streamlines of the ejected material go to infinity those of the current density must be closed and return to the disc where the emf is.

This is actually the reason why jet collimation is a subtle issue [29, 30, 43]. Make a cut at a distance z of a jet and compute the total current flowing inside it, namely $I = \int dr 2\pi r J_z$. If this current is non zero and (for instance) negative, then one might say that the Laplace force will be directed towards

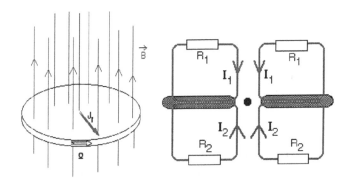

Fig. 1. Left: A rotating disc embedded in a magnetic field induces a current leading to a magnetic braking (Barlow's wheel: see e.g. *http:// www.sparkmuseum.com/MOTORS.HTM* for many illustrations). **Right**: A MAES can be seen as two independent electric circuits, each corresponding to a jet. Asymmetric jets can thus be easily achieved, even with a symmetric poloidal field

the jet axis (Ampere's theorem tells that B_ϕ is negative in that case). This is the basic idea of the "magnetic hoop stress" that provides a self-confinement to jets. However, the local magnetic force is actually $J \wedge B$ and depends on the radial distribution of $J_z(r)$! This depends on the lateral boundary conditions (jet axis and outer edge) but also on what happened upstream (or in the past, if we follow a lagrangian particle): since jet acceleration is a conversion of electric into kinetic power, then jet collimation depends as much on jet acceleration. One cannot therefore solve the jet problem assuming for instance the shape of the field lines: the full MHD equations must be solved.

The current flowing inside jets is precisely the current that allows for accretion. The accretion-ejection phenomenon has therefore to be viewed as a global electric current system.

1.3 Basic Assumptions

Modeling a MAES requires several assumptions:

(1) **Presence of a large scale vertical magnetic field** in the disc. Its origin and amplitude remain an open question. For the purpose of illustration, we will assume a positive vertical component B_z anchored in the disc (bipolar topology).

(2) **Single-fluid MHD**: matter is assumed ionized enough so that all species (ions, neutrals and electrons) are well coupled and can be treated as a single fluid. Such an assumption should always be verified a posteriori for any model but it is seldom made (see e.g. [25] for how to do it).

(3) **Axisymmetry**: using cylindrical coordinates (r, ϕ, z) all quantities are assumed to be independent on ϕ, the jet axis being the vertical axis. Then, $E_\phi = 0$ and all quantities can be decomposed into poloidal (the (r, z) plane) and toroidal components, e.g. $u = u_p + \Omega r e_\phi$ and $B = B_p + B_\phi e_\phi$. A bipolar magnetic configuration can then be described with $B_p = \frac{1}{r} \nabla a \wedge e_\phi$, where the magnetic flux function $a(r, z)$ is an even function of z and with an odd toroidal field $B_\phi(r, -z) = -B_\phi(r, z)$ (Fig. 2).

(4) **Non-relativistic MHD**, since observed motions are non-relativistic (this criterion is enough as long as MHD ordering applies).

(5) **Steady-state**: all astrophysical jets display proper motions and/or emission nodules, showing that they are either prone to some instabilities or that ejection is an intermittent process. However, the time scales involved in all objects (from 1 to 10^2 yrs) are always larger than the orbital time scales in the innermost regions of the underlying accretion disc (close to the star). Therefore, a steady state approach is appropriate as a first step, while numerical simulations will be required to investigate time-dependent flows.

1.4 Governing MHD Equations

According to the aforementioned assumptions, we use the following set of MHD equations (in MKSA units):

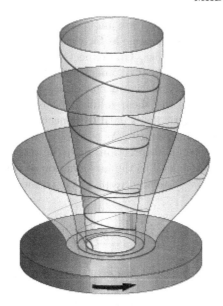

Fig. 2. Axisymmetric jets are made of magnetic surfaces of constant magnetic flux nested around each other and anchored in the disc. Each surface behaves like a funnel whose shape depends on the transfield equilibrium. Solving the jet equations requires to specify several quantities (see text)

Mass Conservation

$$\nabla \cdot \rho \boldsymbol{u} = 0 \tag{1}$$

Momentum Conservation

$$\rho \boldsymbol{u} \cdot \nabla \boldsymbol{u} = -\nabla P - \rho \nabla \Phi_G + \boldsymbol{J} \wedge \boldsymbol{B} + \nabla \cdot \mathsf{T} \tag{2}$$

where Φ_G is the central star gravitational potential and $\boldsymbol{J} = \nabla \wedge \boldsymbol{B}/\mu_o$ is the electric current density. The last term (with the stress tensor T) is actually due to a sustained turbulence inside the disc (it vanishes outside) which allows to transport angular momentum radially in the outward direction (see Terquem's contribution). It is presumably due to the presence of small scale magnetic fields but is usually grossly modeled by an anomalous viscosity $\nu_v = \alpha_v C_s h$, where α_v is a free parameter, C_s the disc sound speed and $h(r)$ the local disc vertical scale height [24, 49].

Ohm's Law and Toroidal Field Induction[1]

$$\eta_m \boldsymbol{J}_\phi = \boldsymbol{u_p} \wedge \boldsymbol{B_p} \tag{3}$$

$$\nabla \cdot \left(\frac{\nu'_m}{r^2}\nabla r B_\phi\right) = \nabla \cdot \frac{1}{r}(B_\phi \boldsymbol{u_p} - \boldsymbol{B_p}\Omega r) , \tag{4}$$

---

[1] See Pelletier's contribution in this volume. Remember that $E_\phi = 0$ while some algebra is required in order to derive (4) from the induction equation.

where $\eta_m = \mu_o \nu_m$ and $\eta'_m = \mu_o \nu'_m$ are anomalous magnetic resistivities. The origin of these resistivities is the same as for viscosity, namely turbulence and they also vanish outside the disc. One expects turbulent media to display anomalous transport effects of heat, momentum but also magnetic flux. Note however that rotation in a Keplerian accretion disc introduces a strong dynamical constraint. Indeed, the shear induced by rotation will unavoidably lead to huge toroidal magnetic fields until reconnection takes place (triggered by e.g. the tearing mode instability). As a consequence the amount of magnetic dissipation in the toroidal direction might be much larger than in the poloidal direction. This has lead to the introduction of two anomalous coefficients, η_m (ν_m) and η'_m (ν'_m), related respectively to the poloidal and toroidal fields [21].

Perfect Gas Law

$$P = \rho \frac{k_B}{\bar{\mu} m_p} T \,, \tag{5}$$

where m_p is the proton mass and $\bar{\mu}$ a generalized "mean molecular weight" (in a fully ionized plasma $\bar{\mu} = 1/2$). This expression assumes that all fluids (electrons, neutrals and ions) have the same temperature T. This is fulfilled only if the thermalization time scale (usually done through collisions) is short enough. Such an assumption should always be verified a posteriori [25].

Energy Equation

$$P\boldsymbol{u} \cdot \nabla \ln \frac{T}{\bar{\mu}} = (\gamma - 1)(Q + P\boldsymbol{u} \cdot \nabla \ln \rho) \,, \tag{6}$$

where $Q = Q^+ - Q^-$ is the sum of all heating Q^+ and cooling Q^- terms (including thermal conduction) and γ the adiabatic index. There are many unsolved issues related to this exact equation for a single fluid.

(1) Inside the disc, turbulence leads to an energy dissipation $Q^+ = \eta_m J_\phi^2 + \eta'_m J_p^2 + \rho \nu_v |r \nabla \Omega|^2$, respectively Joule and "viscous" heating, but also to a cooling due to an energy transport by anomalous thermal conductivity. Moreover, the disc being optically thick, the radiation transport critically depends on the local opacity regime, which varies both with radius and height. Moreover, the disc surface is also the optically thick-thin transition, which is always an issue (see [21] for a discussion). Besides, the energy equation in a standard accretion disc is usually written $Q^+ = Q^-$, the other terms being of the order $(h/r)^2$ [24]. But these terms are important in the jet and cannot be neglected.

(2) In the jet itself, although radiation may not be the dominant cooling term, it must be taken into account if one desires to compute e.g. the jet (forbidden or permitted) emission lines or even radio continuum. A realistic and self-consistent treatment of the energy equation is therefore still out of range (even if one decouples the disc and its jets) and some stratagems must be used.

The simplest way to deal with the energy equation in a MAES (valid in both the disc and its jets) is to use a polytropic equation of state $P = K\rho^\Gamma$, where the polytropic index Γ can be set to vary between 1 (isothermal case) and γ (adiabatic case). Note that K has to vary radially but remains constant along each field line: the jet entropy is thus fixed by the conditions prevailing at the disc surface.

A more sophisticated approach can be done by prescribing the function Q along the field lines (this is equivalent to prescribing a variation of the polytropic index Γ). This will be discussed further in Sect. 3.

2 Physics of Jet Emitting Discs

In this section, I will briefly discuss all the relevant physical effects that have to be covered in order to consistently describe Jet Emitting Discs or JEDs.

2.1 Mass Conservation

The disc accretion rate is defined as $\dot{M}_a = -2\int_0^h 2\pi r\rho u_r dz$. In a standard accretion disc (hereafter SAD) $\dot{M}_a$ is a constant both in time and radius. On the contrary, a JED displays mass loss at its surfaces so that $\dot{M}_a$ must vary with the radius. This mass loss is parametrized by $\dot{M}_a(r) \propto r^\xi$ where the ejection index $\xi > 0$ is a measure of the disc ejection efficiency: the larger ξ the larger the mass loss. The global mass conservation in a JED is $\dot{M}_a(r_e) - 2\dot{M}_j = \dot{M}_a(r_i)$ where r_e and r_i are respectively the outer and inner radii of the JED and $\dot{M}_j$ is the mass flux from one side of the disc (Fig. 3). The ejection to accretion rates ratio is $2\dot{M}_j/\dot{M}_a(r_e) \simeq \xi \ln(r_J/r_i)$ and depends on both the ejection index ξ and the radial extent of the JED (it will be shown later that ξ is smaller than unity). The goal is of course to compute ξ as a function of the disc physical conditions.

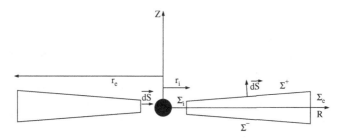

Fig. 3. Sketch of the Jet Emitting Disc (JED) established between r_i and r_e. The surface of the jet is determined by the magnetic surface anchored on r_e. While the inner radius r_i is probably defined by some equilibrium with the stellar magnetosphere (see Sect. 4.6), the outer radius r_e is free (it depends mostly on the magnetic flux available in the disc)

2.2 Poloidal Field Diffusion

Let us assume a smooth flux function a (see Sect. 1.3) so that $a(r,z) \simeq a_o(r)(1 - z^2/2l^2)$ where $l(r)$ is the magnetic flux vertical scale height. Then, the bending of the poloidal field lines is measured at the equatorial plane by the magnetic Reynolds number $\mathcal{R}_m = -ru_r/\nu_m = r^2/l^2$. Such a bending is due to the interplay between advection by the accreting material and the turbulent magnetic diffusivity ν_m. It has been prescribed with $\nu_m = \alpha_m V_A h$, where $V_A = B_z/\sqrt{\mu_o \rho}$ is the Alfvén speed at the disc midplane [20].

Now, magneto-centrifugal acceleration requires field lines bent enough at the disc surface, namely $B_r^+ \geq B_z$ (Blandford & Payne [5], quantities evaluated at $z = h$ are denoted with a superscript "+"). Since $B_r^+/B_z \simeq \mathcal{R}_m h/r$ this implies $\mathcal{R}_m \geq r/h$.

2.3 Angular Momentum Conservation

The disc angular momentum can be transported by two means: (a) radially through a "viscous" turbulent torque which is probably triggered and sustained by an MHD instability such as the magneto-rotational instability (see Terquem's contribution, [2] and references therein); (b) vertically by the jets. The viscous torque writes $F_{visc,\phi} \sim -\alpha_v P/r$ where P is the total (gas+radiation) pressure and α_v the so-called Shakura-Sunyaev parameter. The torque due to the jets writes $F_{mag,\phi} = J_z B_r - J_r B_z$ and its vertical behavior strongly depends on the radial current density J_r. At the disc midplane $F_{mag,\phi} = -J_r B_z \sim B_\phi^+ B_z/\mu_o h$ and the disc angular momentum conservation reads

$$1 + \Lambda \simeq -\frac{ru_r}{\nu_v} = \mathcal{R}_e = \mathcal{R}_m \left(\frac{\nu_m}{\nu_v}\right) , \qquad (7)$$

where ν_v is the turbulent "viscosity" and $\Lambda = F_{mag,\phi}/F_{visc,\phi}$. In a turbulent medium, one usually assumes that all anomalous transport coefficients are of the same magnitude so that $\nu_m \sim \nu_v$. In that situation, one gets the following consequences:
- In a SAD, there is no jets and $\Lambda = 0$. Then $\mathcal{R}_m \simeq \mathcal{R}_e \sim 1$ and, indeed, field lines are too straight for a magneto-centrifugal driving [36];
- In a JED, jets require $\mathcal{R}_m \sim \mathcal{R}_e \geq r/h$ and thus $\Lambda \sim r/h \gg 1$: all the angular momentum must then be carried away by the jets, which results in an accretion velocity much larger than in a SAD. The "viscous" torque is totally negligible (in contrast to what is often assumed, e.g. [42]).

This very important constraint ($\Lambda \geq r/h$) can only be achieved if $-B_\phi^+ B_z/\mu_o \sim P$, that is with equipartition fields [21]. Let us introduce here two important parameters: the disc magnetization $\mu = B_z^2/\mu_o P$ and the magnetic shear $q \simeq -B_\phi^+/B_z$. If $q\mu$ is not close to unity then no magneto-centrifugally driven jets can be launched from accretion discs.

2.4 Toroidal Field Induction

Magnetic driving of jets requires that the magnetic field starts to accelerate material at the disc surface. Hence, a JED must provide a transition from $F_{mag,\phi} < 0$ at $z = 0$ to $F_{mag,\phi} > 0$ at $z = h$ and beyond. The only way to achieve this is by allowing J_r to decrease on a disc scale height. The vertical profile of J_r is provided by the induction (4) which, in a thin accretion disc, writes (see [21])

$$\eta'_m J_r \simeq \eta'_o J_o + r \int_0^z dz \, \boldsymbol{B}_p \cdot \nabla \Omega - B_\phi u_z \, , \tag{8}$$

where $\eta'_o J_o = \eta'_m J_r(z = 0)$. With no differential rotation, J_r would remain constant and so would $F_{mag,\phi}$ (< 0). In order to make J_r decrease on a disc scale height, the disc differential rotation term must balance the term $\eta'_o J_o$, due to the Faraday's induction law (the Barlow's wheel current). This can be done as long as $-B_\phi^+/B_r^+ \sim 1/\alpha_m$ (with $\nu_m \sim \nu'_m$). Using the fact that $B_r^+ \geq B_z$, one gets a magnetic shear $q \geq 1/\alpha_m$.

2.5 Disc Vertical Equilibrium

It is worthwhile to consider the following general equality

$$(\boldsymbol{J} \wedge \boldsymbol{B}_p) \cdot \boldsymbol{B}_\phi = -(\boldsymbol{J}_p \wedge \boldsymbol{B}_\phi) \cdot \boldsymbol{B}_p \, . \tag{9}$$

When the magnetic torque (lhs) is negative, so must be the projection of the Lorentz force on the poloidal field (rhs). Thus, deep within the disc, the poloidal Lorentz force is directed outwardly and towards the disc midplane. A quasi MHS equilibrium is therefore established with the balance between the total (gas+radiation) pressure gradient on one side and the magnetic compression due to the radial and toroidal field components and the gravity on the other side. Now, as one goes up in z and the magnetic torque changes its sign, the disc material starts to be azimuthally accelerated. Correspondingly, the projection of the Lorentz force becomes also positive and helps to lift material out of the disc.

This can be done in two ways (Fig. 4): (a) with a negative vertical component of the Lorentz force but a large radial component; (b) with a positive vertical component and a smaller negative radial component. Case (a) corresponds to a small mass flux ($\xi < 1/2$) where disc material must be lifted against the magnetic compression by the sole effect of the (gas+radiation) pressure gradient. Case (b) leads to a large mass flux ($\xi > 1/2$) because of the magnetic pull due to the toroidal field pressure.

In fact, it can be shown analytically that only solutions with $\xi < 1/2$ can be stationary: solutions with large mass fluxes do not have enough power to allow for super-Alfvénic jets [16]. This has an important consequence on disc physics. Since $B_r^+ \geq B_z$, the total pressure gradient can overcome the

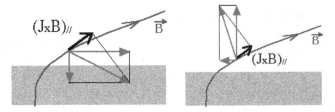

Fig. 4. Magnetic acceleration arises whenever the projection of the Lorentz force on a poloidal field line becomes positive. This can be achieved in two ways, either with a downward vertical magnetic compression or a strong outward pressure force due to the toroidal field. The former leads to a small ejection efficiency and has current lines coming out of the disc surface ($J_z > 0$) and entering at the inner radius. The latter has a strong ejection efficiency with the current entering the disc at its surfaces ($J_z < 0$). Only small ejection efficiencies allow for steady state solutions [16]

magnetic compression due to B_r only if μ is not larger than unity. The same constraint holds for the toroidal field which implies that α_m must be of the order of unity. Finally, using the fact that $\Lambda \sim r/h \gg 1$ in a JED, one obtains that μ cannot be too small. Thus, the parameter space for a JED is $\mu \sim q \sim \alpha_m \sim 1$. Note also that a JED is thinner than a SAD because of the additional magnetic compression [21, 56].

2.6 Disc Radial Equilibrium

The quasi MHS radial equilibrium leads to an angular velocity

$$\Omega^2 = \Omega_K^2 \left(1 + \frac{\partial P/\partial r}{\rho \Omega_K^2 r} - \frac{(\boldsymbol{J} \wedge \boldsymbol{B})_r}{\rho \Omega_K^2 r} + \frac{u_r^2}{\Omega_K^2 r^2} \frac{\partial \ln u_r}{\partial \ln r} \right) . \tag{10}$$

The deviation to the Keplerian rotation law $\Omega_K = \sqrt{GM/r^3}$ due to the radial (gas+radiation) pressure gradient is roughly of order $(h/r)^2$ at each altitude. This is because P scales with the density, which is not the case of the radial magnetic tension. At the disc midplane, it causes a deviation which is of the order $\sim \mu \mathcal{R}_m (h/r)^2 \sim h/r$ but increasing as $1/\rho$. Thus, thin accretion discs with $h/r \ll 1$ will be mostly rotating at (sub-) Keplerian speeds but a problem arises when $h \sim r$ (as in ADAFs Narayan et al. [41] or in self-gravitating discs, for instance). Indeed, it would imply a negative rhs at the disc surface which certainly means that no steady-state accretion-ejection solution can be found in that case. Note however that there is a priori no reason to ever have $h \sim r$ in JEDs: they are colder (see below) and more squeezed (see above) than a SAD.

2.7 Energy Budget

The global energy budget is obtained by applying the energy conservation equation to the whole volume occupied by the JED. This equation writes

$P_{acc} = 2P_{MHD} + 2P_{rad}$ where

$$P_{acc} \simeq \frac{GM\dot{M}_a(r_e)}{2r_i} \qquad (11)$$

is the mechanical power liberated by the accreting material between r_e and r_i, $P_{MHD} = \int \boldsymbol{S}_{MHD} \cdot d\boldsymbol{S}$ is the flux through one disc surface of the MHD Poynting vector $\boldsymbol{S}_{MHD} = \boldsymbol{E} \wedge \boldsymbol{B}/\mu_o \simeq -\Omega_K r B_\phi \boldsymbol{B}_p/\mu_o$. So, energy conservation in a JED tells us that the available accretion power is shared between a flux of electromagnetic energy powering the jets and radiation due to heat dissipation within the disc ($2P_{rad} = P_{diss}$). This dissipation is due to the fact that in a disc where turbulent magnetic diffusivity/resistivity and viscosity are assumed, there is always some heat production. The simplest and crudest way to estimate this dissipation is using effective transport coefficients so that it writes $P_{diss} = \int_V dV (\eta_m J_\phi^2 + \eta_m' J_p^2 + \eta_v (r\partial\Omega/\partial r)^2)$, namely Joule and "viscous" heating.

Since the "viscous" torque is negligible with respect to the jet torque, only a small fraction of the energy will be dissipated by viscosity. On the other hand, the most interesting aspect of angular momentum removal by jets is that the associated Joule dissipation implies also only a small fraction of the available energy. As a consequence, most of the liberated accretion power goes into the jets [20, 21]! Precisely, this can be written as

$$\frac{2P_{MHD}}{P_{acc}} \simeq \frac{\Lambda}{1+\Lambda} \quad \text{and} \quad \frac{2P_{rad}}{P_{acc}} \simeq \frac{1}{1+\Lambda}, \qquad (12)$$

where the ratio of the jet to the viscous torque $\Lambda \sim r/h \gg 1$. This property of JEDs has two important consequences: (i) the disc itself being weakly dissipative, it may well be unobservable leading to the (wrong) idea that there is no disc; (ii) a JED is cooler than a SAD fed with the same accretion rate, which leads to a smaller aspect ratio h/r.

2.8 Links Between Jet and Disc Physics

The previous sections showed the crucial role played by the magnetic diffusivity within the turbulent JED. On the contrary, jets are best described by an ideal MHD formalism ($\nu_m = \nu_m' = 0$). This leads to the existence of 5 invariants along each magnetic surface for polytropic jets[2]. The Bernoulli equation is obtained by projecting the momentum (2) along $\boldsymbol{B}_p$ whereas the transfield or Grad-Shafranov equation by projecting it along ∇a (perpendicular to $\boldsymbol{B}_p$). For more details see Tsinganos' contribution (this volume).

MHD simulations of jets driven by accretion discs usually assume magnetic field lines rotating at Keplerian speeds and negligible enthalpy leaving

---

[2] The magnetic surface rotation rate $\Omega(a)$, the mass to magnetic flux ratio $\eta(a)$, the total specific angular momentum $L(a)$, energy $E(a)$ and entropy $K(a)$.

therefore 3 free and independent boundary conditions to be specified at each radius (see e.g. [1, 46] and references therein). These are often the density $\rho(r)$, vertical velocity $u_z(r)$ and magnetic field $B_z(r)$ distributions. However, the study of MAES shows that not all distributions allow for steady state jets: there is a strong interplay between the disc and its jets. Such an interplay appears in the form of analytical links between jet invariants and parameters describing the disc. These links can be found in Casse & Ferreira [8]; Ferreira [17] and Ferreira & Casse [18]. Of all disc parameters the disc ejection efficiency ξ plays a major role. Indeed, the knowledge of ξ allows to define almost all jet properties. But in order to obtain the allowed values for ξ, the full set of MHD equations must be solved.

3 A Glimpse on Self-similar Solutions

3.1 Mathematical Method

This is done by a separation method allowing to transform the set of partial differential equations (PDE) into two sets of ordinary differential equations (ODE) with singularities. Now, the gravitational potential in cylindrical coordinates is

$$\Phi_G(r, z) = -\frac{GM}{r}\left(1 + \frac{z^2}{r^2}\right)^{-1/2} \tag{13}$$

and it is expected to be the leading energy source and force in accretion discs. Thus, if JEDs are settled on a large range of radii (so that we do not care about the radial inner and outer boundaries), then the magnetic energy density has to follow gravity in order to match it everywhere. It is therefore justified to look for solutions of the form $A(r, z) = G_A(r)f_A(\frac{z}{r})$ for any physical quantity $A(r, z)$. Moreover, since gravity is a power law of the disc radius, we will use the *self-similar* Ansatz $A(r, z) = A_e\left(\frac{r}{r_e}\right)^{\alpha_A} f_A(x)$ where $x = z/h(r)$ is our self-similar variable with $h \propto r$ and r_e is the JED outer radius. Because all quantities have power law dependencies, the resolution of the "radial" set of equations is trivial and provides algebraic relations between all exponents. The most general set of radial exponents allowing to take into account **all** terms in the dynamical equations leads to the following important constraint

$$\beta = \frac{3}{4} + \frac{\xi}{2}, \tag{14}$$

where the magnetic flux distribution writes $a(r) \propto r^\beta$. As an illustration, the solutions obtained by Blandford & Payne [5] used $\beta = 3/4$, i.e. $\xi = 0$. In general, all self-similar models of disc driven jets not addressing the disc dynamics use a magnetic field distribution inconsistent with the jet mass loading [5, 12, 13, 44, 55].

All quantities $f_A(x)$ are obtained by solving a system of ODE which can be put into the form

$$\begin{pmatrix} & & \\ \cdots & \mathbf{M} & \\ & & \cdots \end{pmatrix} \cdot \begin{pmatrix} \frac{df_1}{dx} \\ \vdots \\ \frac{df_n}{dx} \end{pmatrix} = \begin{pmatrix} \cdots \\ \mathbf{P} \\ \cdots \end{pmatrix},$$

where $\mathbf{M}$ is a 8x8 matrix in resistive MHD regime, 6x6 in ideal MHD [21]. A solution is therefore available whenever the matrix $\mathbf{M}$ is inversible, namely its determinant is non-zero. Starting in resistive MHD regime, $det\,\mathbf{M} = 0$ whenever $V^2(V^2 - C_s^2) = 0$ where C_s is the sound speed and $V \equiv \boldsymbol{u} \cdot \boldsymbol{n}$ is the critical velocity. The vector $\boldsymbol{n} = (\boldsymbol{e}_z - \frac{z}{r}\boldsymbol{e}_r)(1 + \frac{z^2}{r^2})$ provides the direction of propagation of the only waves consistent with an axisymmetric, self-similar description (see Tsinganos' contribution). Therefore, close to the disc, the critical velocity is $V \simeq u_z$, whereas far from the disc it becomes $V \simeq u_r$ (no critical point in the azimuthal direction). Inside the resistive disc, the anomalous magnetic resistivity produces such a dissipation that the magnetic force does not act as a restoring force and the only relevant waves are sonic. Note also that the equatorial plane where $V = 0$ is also a critical point (of nodal type since all the solutions must pass through it). This introduces a small difficulty as one must start the integration slightly above $z = 0$. In the ideal MHD region, $det\,\mathbf{M} = 0$ whenever $(V^2 - V_{SM}^2)(V^2 - V_{FM}^2)(V^2 - V_{An}^2)^2 = 0$, namely when the flow velocity V successively reaches the three phase speeds V_{SM}, V_{An} and V_{FM}, corresponding respectively to the slow magnetosonic (SM), Alfvén and fast magnetosonic (FM) waves. The phase speeds of the two magnetosonic modes are $V_{SM,FM}^2 = \frac{1}{2}\left(C_s^2 + V_{At}^2 \mp \sqrt{(C_s^2 + V_{At}^2)^2 - 4C_s^2 V_{An}^2}\right)$ where V_{At} is the total Alfvén speed and $V_{An} = \boldsymbol{V}_{Ap} \cdot \boldsymbol{n}$. These expressions are slightly modified by the self-similar ansatz. Note however that the condition $V = V_{An}$ is equivalent to $u_p = V_{Ap}$.

How do we proceed ? We fix the values of the four disc parameters $(\varepsilon = h/r, \alpha_m = \nu_m/V_A h, \chi_m = \nu_m/\nu_m', \mathcal{P}_m = \nu_v/\nu_m)$ and some guesses for the disc magnetization μ and ejection efficiency ξ. Starting slightly above the disc midplane where all quantities are now known, we propagate the resistive set of equations using a Stoer-Burlisch solver for stiff equations. As $x = z/h$ increases, the flow reaches an ideal MHD regime and we shift to the corresponding set of equations. Care must be taken in order to not introduce jumps in the solution while doing so. The smooth crossing of the SM point can only be done with a critical value for μ. We thus modify our initial guess until the solution gets close enough to the critical point and jump across it (leapfrog method). The same must be done for the Alfvén point which requires a critical value for ξ. Each time another guess for ξ is made, one has to find again the corresponding critical value for μ (Fig. 5). The crossing of the last critical point (FM) does not bring much more information on MAES physics and will be discussed in Sect. 4.

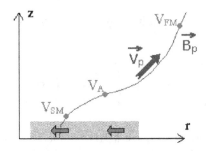

Fig. 5. Once material has left the resistive MHD zone, it is frozen in a particular field line and encounters the three MHD critical points. The smooth transition between resistive and ideal MHD regimes already selects the MAES parameter space (see Sect. 2)

3.2 Typical Solutions

Only the most salient features of self-similar accretion-ejection solutions will be discussed here (see [17] and [18]).

Cold Solutions

Cold solutions are defined here by an isothermal [16, 21] or adiabatic [8] energy equation. Since the plasma pressure P is $(h/r)^2$ smaller than the gravitational energy density, such an energy equation ensures that the jet enthalpy is negligible with respect to gravity and magnetic fields [5].

Figure 6 shows the velocity components in both the JED and the jets as a function of the self-similar variable x, *along a magnetic surface* for typical solutions with $h/r = 0.01$ but different ejection efficiencies ξ. The disc surface is located at $x = 1$ and the Alfvén point is reached at $x \sim 100$ ($z_A \sim r_A$). Note that the disc vertical velocity is negative within the disc (material is falling) and becomes positive only slightly before the point where the radial velocity itself becomes positive. This happens roughly at the disc surface (see bottom right panel) but still in the resistive MHD regime. The SM point is crossed at $x \simeq 1.6$. All velocity components are comparable at the Alfvén point (this also holds for the magnetic field). Beyond that point, the plasma inertia overcomes the magnetic tension and the magnetic surface opens tremendously. This leads to the build-up of a sheared magnetic configuration (the ratio $|B_\phi/B_p|$ increases). Note that the structure of the jet can be characterized by two families of intertwined helices: the plasma streamlines (wound in the same direction as the disc rotation) and the magnetic field lines (wound in the opposite direction).

The magnetic acceleration is so efficient that all available MHD energy is transferred into jet kinetic energy. From Bernoulli equation one gets analytically the asymptotic jet velocity $v_j = \Omega_o r_o \sqrt{2\lambda - 3}$ where $\Omega_o r_o$ is the

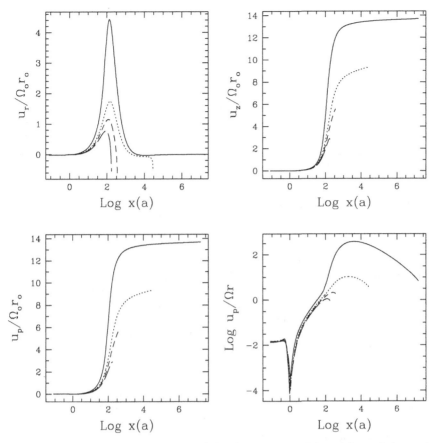

Fig. 6. Components of the jet poloidal velocity u_p and logarithm of the ratio of the poloidal to the azimuthal velocity, measured along a magnetic surface for $\xi = 0.005$ (*solid line*), 0.01 (*dotted line*), 0.02 (*short-dashed line*) and 0.05 (*long-dashed line*) ($\varepsilon = 10^{-2}$, $\alpha_m = 1$). For these typical cold solutions, the jet always reaches its maximum velocity, mainly as a vertical component (the jet opening angle is $\tan\theta = B_r/B_z = u_r/u_z$). Inside the disc, matter is being accreted with a velocity of order ε the Keplerian velocity $\Omega_o r_o$ [16]

Keplerian speed at the jet footpoint r_o and λ is the magnetic lever arm parameter [5]. This important jet parameter is actually related to the disc ejection efficiency $\lambda \simeq 1 + 1/2\xi$ [16].

The disc parameter space has been thoroughly investigated for cold solutions. It is very narrow with typical values $\xi \sim 0.01$ and $0.1 < \mu < 1$, with the following approximate scaling

$$\xi \sim 0.1\,\mu^3 \qquad (15)$$

Although its validity holds only in a quite narrow interval, it shows that *the stronger the field the more mass is ejected.* No solution has been found outside

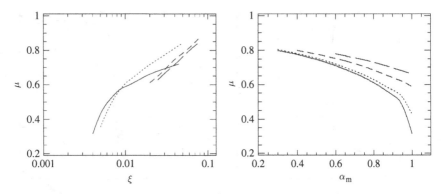

Fig. 7. Disc parameter space for isothermal jets [16] (adiabatic jets present no qualitative difference, see Casse & Ferreira [8]). **Left:** disc magnetization μ as a function of the disc ejection efficiency ξ for $\alpha_m = 1$ and various disc aspect ratios: $\varepsilon = h/r = 10^{-1}$ (*solid line*), 10^{-2} (*dotted line*), 10^{-3} (*short-dashed line*) and $7\ 10^{-4}$ (*long-dashed line*). The main effect of decreasing ε is to shift the range of allowed ξ to higher values (but with a more limited range). **Right:** Influence of the turbulence parameter α_m on the disc magnetization μ for $\varepsilon = 10^{-1}$ and various ejection efficiencies: $\xi = 0.004$ (*solid line*), 0.005 (*dotted line*), 0.01 (*short-dashed line*) and 0.02 (*long-dashed line*). The minimum level of MHD turbulence is limited by the value of the induced toroidal field allowing trans-Alfvénic jets, whereas the maximum level has been arbitrarily fixed to unity

the range $0.0007 < \varepsilon = h/r < 0.3$ and $0.3 < \alpha_m < 3$ (Fig. 7). As pointed out previously, there is no solution with a dominant viscous torque. All solutions exhibit a high degree of collimation: actually, they even undergo recollimation towards the axis which should result in a shock [16]. However, the subsequent behaviour of the jet after that shock cannot be treated within self-similarity.

Warm Solutions

Warm solutions are obtained by solving (6) with a prescribed self-similar function Q. Several physical effects can be simulated that way:

- Heat deposition at the disc surface only: the function Q reaches a maximum at the disc upper layers and then decreases rapidly (to recover adiabatic jets). This mimics the effect of disc illumination by stellar UV and X rays. Alternatively, this energy could arise from the dissipation of a small fraction of the accretion energy, released in these layers by turbulence. Remarkably the mass load can be significantly enhanced, with ejection efficiencies up to $\xi \simeq 0.46$ [9].
- Heating of the sub-Alfvénic regions: the function Q is non zero in these regions only with subsequent adiabatic or polytropic jets. This mimics the effect of some "coronal" heating as in the solar wind or, alternatively, the pressure due to an inner flow (e.g. stellar or magnetospheric wind)

ramming into the disc wind. Under some circumstances, the field lines are forced to open much more than they would which results in a different jet dynamical behaviour. In particular, self-similar jets can smoothly cross the last modified FM critical point [18, 55]. See Fig. 8 for an example.

• Heating of the whole jet: this has not yet been done in the framework of disc driven jets (but could easily be done by assuming a positive function Q everywhere in the jet). The reason is that such jets would not be significantly modified by a warmer material (in contrast with stellar winds). However, this is interesting for comparing models to observations. Indeed, observed jets display temperatures of some 10^4 K that require some heating mechanism(s) overcoming the huge cooling due to the jet expansion (so called adiabatic cooling). It has been shown that ambipolar diffusion is not enough and that some turbulent or shock heating must be at work [25].

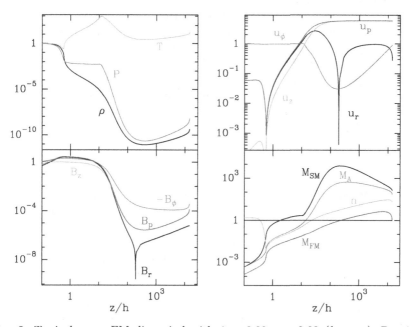

Fig. 8. Typical super-FM disc wind with $\xi = 0.03, \epsilon = 0.03$ ($h = \epsilon r$). Density, pressure and temperature are normalized to their value at the disc midplane, the magnetic field components to $B_z(z = 0)$ and the velocities to the Keplerian speed at the anchoring radius r_o. All magnetic field components remain comparable from the disk surface to the Alfvén point. Note that the density profile inside the disc, where both u_r and u_z are negative, is very different from a gaussian. Recollimation takes place at $z \simeq 3 \ 10^3 r_o$. The lower right panel shows the various critical Mach numbers (e.g. $M_{SM} = V/V_{SM}$) appearing in the self-similar equations. The usual fast Mach number, $n = u_p/u_{FM}$, becomes greater than unity much sooner than the critical one $M_{FM} = V/V_{FM}$ [18]

4 Concluding Remarks

4.1 What's Next?

The theory of *steady* jet production from Keplerian accretion discs is now completed. The physical conditions required to thermo-magnetically drive jets are known, all relevant physical processes have been included in the framework of mean field dynamics. Of course, there are still many unsolved questions:

(i) Can a sustained MHD turbulence maintain $\alpha_m \sim 1$? This is a huge constraint that deserves a thorough investigation.

(ii) Observations of T Tauri jets favor solutions with large ejection efficiencies ($\xi \sim 0.1$, Pesenti et al. [48]) requiring additional heating at the disc surface. A theoretical assessment of this heating must be undertaken.

(iii) What is the stability of MAES? As will be seen below, there was some claims that MAES were unstable but they were proven to be wrong. On the other hand, jets do show time dependent features and one must clearly go beyond steady state models. On that respect, numerical simulations will be very helpful.

(iv) Disc driven winds do not treat the star-disc interaction. Understanding the whole process of star formation requires now to address this crucial issue as it pinpoints the problem of the stellar angular momentum removal. This is further discussed below.

4.2 Biases of Self-similarity

Self-similarity allows to take into account all dynamical terms in the equations and, as such, is the best means to solve in a self-consistent way the steady-state accretion-ejection problem. However, there is a price to pay...

   (i) The asymptotic behaviour is obviously biased since, for instance, neither inner nor outer pressures can be taken into account. In fact, no realistic "radial" boundary condition whatsoever can be dealt with. When modeling an astrophysical jet, this implies for instance to truncate the solution at one inner and outer radius. But there is another aspect, less known and more subtle. Contopoulos & Lovelace [13] and Ostriker [44] obtained jet solutions within the same self-similar framework but with different asymptotic behaviors. The reason stems from the fact that they played around with β (flux function $a(r) \propto r^\beta$) as if it were a free function whereas the mathematical matching with a Keplerian disc imposes its value. On the other hand, Pelletier & Pudritz [47] obtained also recollimating *non self-similar* solutions, which indicates that recollimation can indeed be physical and not entirely due to self-similarity. In fact, it can be shown that recollimation of a jet launched from a Keplerian accretion disc is possible whenever the radial profile of the ejection efficiency ξ is smooth enough [16].

(ii) The regularity conditions are to be imposed at the modified points and not at the usual ones (see Tsinganos, this volume). However, these locations coincide for both the slow (SM) and the Alfvén points so that one can be confident that there is no bias there. However, this is not so for the fast magnetosonic point. Self-similar trans-FM solutions require an Alfvén surface very close to the disc [18], which can only be done by the action of a large pressure in the sub-Alfvénic region. This is obviously a strong bias since it is not clear whether such a pressure is indeed provided in astrophysical objects. Note however that crossing this modified FM point is more a theoretician satisfaction than anything else: it gives no additional physical insight on the disc physics.

(iii) The local disc physical conditions as obtained with self-similar solutions are not biased. The physical processes are well identified and understood and can be sometimes even obtained in a pure analytical manner. They have been also confirmed by numerical experiments of Casse & Keppens [10, 11]; Zanni et al. [57] (although one might object that numerical experiments were actually tested with the help of semi-analytical solutions). Figure 9 shows the result of such a simulation, where the jet mass loading is computed in a consistent way.

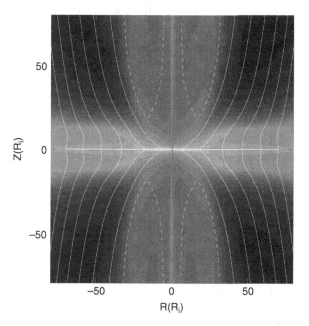

Fig. 9. Poloidal snapshot of an accretion disc driving self-confined jets. The grey background is the density contours, the white lines are the magnetic field lines. The outflow becomes super fast magnetosonic and the overall system reaches a quasi-steady state similar to those obtained in the self-similar models of Ferreira et al. Such a simulation was possible thanks to the inclusion of an anomalous magnetic resistivity and an equipartition vertical magnetic field. Taken from [10]

4.3 Is Accretion-ejection Unstable?

There has been some claims in the literature that the accretion-ejection process itself was unstable [7, 37]. The idea was the following. Start from a steady picture where the accretion velocity u_r at the disc midplane is due to the jet torque. It leads to a bending of the poloidal field lines described by an angle θ with the vertical. Now imagine a small perturbation δu_r enhancing the accretion velocity. Then, according to these authors, the field lines would be more bent (θ increases) which would lead to lower the altitude of the sonic point. Because the sonic point would be located deeper in the disc atmosphere, where the density is higher, more mass would be henceforth ejected which would then increase the total angular momentum carried away by the jet. This means that the torque due to the jet is enhanced and will, in turn, act to increase the accretion velocity. Thus, the accretion-ejection process is inherently unstable.

The whole idea of this instability is based upon a crude approximation of the disc vertical equilibrium. In fact, the magnetic field produces a strong vertical compression so that, as θ is increased, *less* mass is being ejected, not more. This has been pointed out by Königl & Wardle [33] and Königl [31] and is indeed verified in the full MAES calculations reported here.

4.4 Magnetic Fields in Accretion Discs

The necessary condition for launching a self-collimated jet from a Keplerian accretion disc is the presence of a large scale vertical magnetic field close to equipartition [21], namely

$$B_z \simeq 0.2 \left(\frac{M}{M_\odot} \right)^{1/4} \left(\frac{\dot{M}_a}{10^{-7} M_\odot/yr} \right)^{1/2} \left(\frac{r_o}{1 \text{ AU}} \right)^{-5/4+\xi/2} \text{ G} , \qquad (16)$$

This value is far smaller than the one estimated from the interstellar magnetic field, assuming either ideal MHD or $B \propto n^{1/2}$ [3, 4, 28]. This implies some decoupling between the infalling/accreting material and the magnetic field in order to get rid off this field. This issue is still under debate. The question is therefore whether accretion discs can build up their own large scale magnetic field (dynamo) or if they can drag in and amplify the interstellar magnetic field? Although no large scale fields have been provided by a self-consistent disc dynamo, this scenario cannot be excluded. But the latter scenario (advection) seems a bit more natural.

Let us assume that the disc material is always ionized enough to allow for some coupling with the magnetic field (and use MHD). The outer parts of the accretion disc will probably take the form of a SAD with no jets and almost straight ($\mathcal{R}_m \sim 1$) field lines [36]. In that case, the steady-state solution of the induction equation for the poloidal field is $B_z \propto r^{-\mathcal{R}_m}$ [23]. Hence, as a result of both advection and (turbulent) diffusion, the magnetic field in a SAD will be a power-law of the radius.

Can a SAD transport B_z and allow for a transition to an inner JED? This will be so if there is some transition radius (the outer JED radius r_e) where $\mu = B_z^2/\mu_o P$ becomes of order unity. In a SAD the total pressure writes $P = \frac{\dot{M}_a \Omega_K^2 h}{6\pi\nu_v} \propto r^{-3/2-\delta}$ with $h(r) \propto r^\delta$. Since δ is always close to unity in circumstellar discs, one gets $\mu \propto r^{-\epsilon}$ with $\epsilon \sim 1$. Thus, it can be readily seen that it is indeed reasonable to expect such a transition (computing it is another matter), at least in some objects. The recent Zeeman observation of a magnetic field in the accretion disc of FU Or supports this conclusion [14].

4.5 X-winds and Disc Winds

The X-wind model [40, 50, 51, 52, 53] is a rich and complex model but, contrary to common belief, it is an accretion-powered wind launched from the accretion disc. In practice, if the amount of magnetic flux threading the disc is large so that $r_e \gg r_i$, then one gets an "extended disc wind", whereas if the magnetic flux is tiny with $r_e \geq r_i$, one gets an "X-wind" (Fig. 10). The dynamics and asymptotic behaviour of jets will differ strongly between an extended disc wind and an X-wind and can thereby be tested against observations [19]. But this difference arises mainly because of the restricted range in radii in the X-wind case, not because the underlying disc physics is different. The basic phenomena described in Sect. 2 apply as well for the portion of the disc launching the X-wind. Thus, equipartition fields are required, the "viscous" torque is negligible with respect to the jet torque and the angular

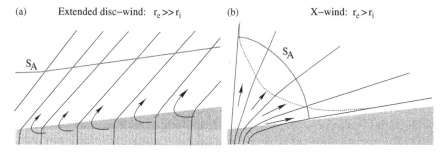

Fig. 10. Two classes of stationary accretion powered disc winds. (**a**) "extended disc winds", when the magnetic flux threading the disc is large enough so that a large radial extension of the whole accretion disc drives jets ($r_e \gg r_i$). The Alfvén surface S_A is expected to adopt a rather conical shape. (**b**) "X-winds", when the magnetic flux is small and only a tiny disc region is driving jets. The Alfvén surface can be either convex or concave, although the latter is probably more physical (since less material can be ejected at the two extremes and the Alfvén point is rejected to infinity). Adapted from Ferreira et al. [19]

momentum carried away by the X-wind is *exactly* the same amount lost by the accreting material. As a consequence, X-winds cannot take away any angular momentum from the central star.

Published material on the dynamics of X-winds contains: (i) a scenario for the origin of B_z (stellar) and the star-disc interaction (leading to the opening of some magnetic field lines); (ii) the calculation of the sub-Alfvénic ideal MHD jet (elliptic domain defined by prescribed boundary conditions); (iii) a somewhat mysterious "interpolation"to a simple jet asymptotic solution. The following questions remain therefore to be addressed:

(1) Can the disc afford the imposed mass flux and field geometry? Indeed, the assumed ejection to accretion mass flux ratio of 1/3 from such a tiny region is huge and would require a fantastic ejection efficiency (ξ of order unity or larger). The calculations of JEDs showed that this is unfeasible in a steady way. However, the huge magnetic field gradients required in the X-wind launching region provide a significantly different situation. This has never been analyzed.

(2) How good is the transfield equilibrium satisfied? There is no mathematical procedure to find a solution of mixed type (elliptic-hyperbolic) PDEs when the singular surfaces are unknown. The trick used for X-winds provides an incomplete solution, but there is maybe some means to fulfill the transverse equilibrium by using an iterative scheme. In any case, this important point is missing in the current published material.

4.6 Magnetic Star-disc Interactions

Nowadays it seems accepted that a lot if not all young stars have a magnetospheric interaction with their circumstellar accretion disc (see Alencar's contribution, this volume). If one assumes that the disc is threaded by a large scale magnetic field, then the question of how this field is connected to the stellar field arises. This is a complex topic that requires the use of numerical experiments but this will not be addressed here as this is the topic of the 5th JETSET school (see however [15, 35] and references therein). Only simple aspects will be discussed here.

First ideas are always simple and so is the stellar magnetic field, assumed up to now to be dipolar and axisymmetric (see Mohanty et al. 2006). We define here the magnetopause as the radius r_m below which all field lines threading the disc are tied to the star whereas beyond r_m, they are disconnected from the star.

The case envisioned within the X-wind scenario assumes a stellar magnetic moment anti-parallel to the disc magnetic field [52]. As a consequence, a neutral surface (where $B = 0$) appears above each disc surface, illustrated by a limiting poloidal field with a Y shape (Fig. 11). The other case, a stellar magnetic moment parallel to the disc magnetic field, has been proposed by Ferreira et al. [22]. The two fields then cancel each other at the disc midplane, defining

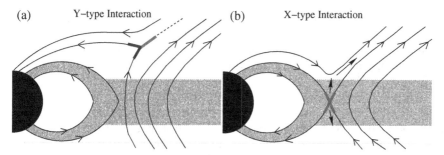

Fig. 11. Two simple axisymmetric star-disc magnetospheric interactions. (a) "Y-type" interaction obtained when the stellar magnetic moment is anti-parallel to the disc magnetic field. A current sheet is formed at the interface between the open stellar field and the disc field. Such a configuration cannot produce per se a wind. (b) "X-type" configuration obtained when the stellar magnetic moment is parallel to the disc magnetic field. A magnetic X-point is generated at the disc midplane where the two fields cancel each other. Unsteady ejection ("ReX-winds") can be launched above this reconnection site. Adapted from Ferreira et al. [19]

a neutral line at a radius r_X where reconnection takes place. This configuration gives rise to "Reconnection X-winds" (hereafter ReX-winds) specifically above this zone.

Accretion Curtains

The first question here is can these simple topologies allow for accretion below[3] r_m? Disc material will accrete only if it looses angular momentum and this depends on both turbulence and the magnetic torque due to the magnetosphere. The magnetosphere will try to make the disc corotate with the protostar so the sign of the torque depends on their relative angular velocity. The corotation radius, $r_{co} = (GM/\Omega_*^2)^{1/3}$, is defined as the radius where the stellar angular velocity Ω_* is equal to the Keplerian one. This gives an estimate of the real angular velocity of the disc (since the disc magnetic field introduces already a deviation). Roughly speaking, if $r_m > r_{co}$ the star rotates faster than the disc and deposits its angular momentum, whereas if $r_m < r_{co}$, the star rotates slower and thus spins down the disc. Note that r_m denotes roughly the radius where the stellar magnetic field becomes dynamically dominant, namely $\mu > 1$. Thus, unless a very efficient turbulent mechanism[4] is operating and transports radially the stellar angular momentum, no accretion is possible when $r_m > r_{co}$ (although such a "propeller" regime is favorable for ejection).

---

[3] Accretion is realized beyond r_m by e.g. the jet torque within the JED and, farther away in the SAD, by the turbulent "viscous" torque.

[4] Note that it should be operating when $\mu > 1$, while the magneto-rotational instability is already quenched at $\mu \sim 1$ [2].

As a consequence, both X-type and Y-type interactions allow for a magnetospheric accretion as long as $r_m < r_{co}$. It is interesting to note that both configurations require an equatorial reconnection zone (interesting for sudden energy dissipation and chondrules, Shu et al. [54], Gounelle et al. [26]). In the case of a Y-type interaction, it arises because of the requirement that the magnetospheric field makes an angle with the vertical large enough in order to allow the disc material to flow inwards. This assumption implies a magnetic neutral "belt" at the disc midplane (see Fig. 1 in Ostriker & Shu [45]), but whose origin and dynamics were not discussed and remain therefore major unsolved issues. In the case of an X-type interaction, the presence of the magnetic neutral line is due to and maintained by the cancellation of the two fields (see Fig. 11). The accreting disc material can cross the resistive MHD region and is lifted vertically by the strong Lorentz force above the reconnection site. The transition from an accretion disc to accretion curtains can be quite smooth in that case.

Stellar Spin Down

The second question is the issue of the stellar angular momentum removal by winds (see Matt & Pudritz [38] for more details and the necessity of winds). As explained earlier, X-winds carry away the angular momentum of the accreting disc material. Thus, such a configuration cannot brake down the protostar (as initially claimed). On the contrary, the X-type configuration provides a very

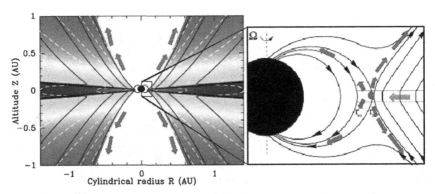

Fig. 12. The ReX-wind configuration [22]. A MAES is established around a protostar whose magnetic moment is parallel to the disc magnetic field. This is a natural situation if both fields (*disc and stellar*) have the same origin. **Left:** black solid lines are streamlines, white dashed lines are contours of equal total velocity (mainly rotation inside the disc) and the background color scale shows the density stratification. The ReX-wind (*arrows*) would be confined and channeled by the outer disc wind. **Right:** sketch of the magnetic configuration leading to Rex-winds and accretion curtains around the magnetic neutral line at r_X. Arrows show the expected time-dependent plasma motion

efficient means to do it [22]. The reason is the possibility to launch disc material above the reconnection site. The scenario is the following (Fig. 12). A stationary extended JED is settled in the innermost regions of the accretion disc and provides open magnetic flux to the star. This magnetic field reconnects at r_X with closed stellar field lines: the disc field contributes thereby to transform closed magnetospheric flux into open flux. At the reconnection site, the disc material is lifted vertically and loaded onto these newly opened field lines, tied to the rotating star. Whenever $r_X > r_{co}$, the star is rotating faster than the loaded material and it undergoes a strong magneto-centrifugal acceleration. This gives rise to the so-called ReX-wind, whose energy and angular momentum are those of the star. Using a toy-model for the magnetic interaction (namely mass, energy and angular momentum conservation

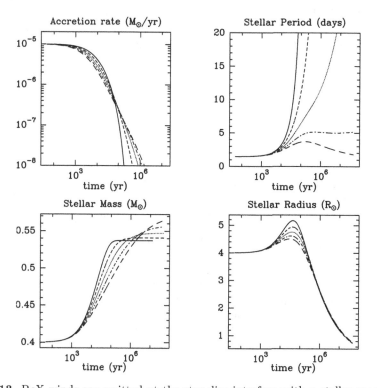

Fig. 13. ReX-winds are emitted at the star-disc interface with a stellar magnetic field varying as $B_* \propto r^{-n}$ (modified dipole). Here, the self-consistent time evolution of the disc accretion rate, protostellar period, mass and radius are shown as functions of n for an ejection to accretion rate ratio $f = 0.1$ and magnetic lever arm $\lambda = 3$ (see, Ferreira et al. [22] for more details). *Solid line*: $n = 3$, *dashed*: $n = 3.41$, *dotted*: $n = 3.87$, *dash-dotted*: $n = 4.4$ and *long-dashed*: $n = 5$. The best solution (stellar period of 8 days after $\sim 10^5$ yrs) is obtained for a compressed dipole (*dotted line*). The initial conditions are a Class 0 protostar of $R_{*,0} = 4R_{sun}$, $M_{*,0} = 0.4M_{sun}$, $T_* = 3000$ K at break-up velocity and with $\dot{M}_{a,0} = 10^{-5}\ M_{sun}yr^{-1}$

of the protostar) Ferreira et al. [22] showed that such winds could brake down a *contracting* protostar (see Fig. 13) on time scales that are comparable to the duration of the embedded phase (Class 0 and I sources). The protostar was assumed to rotate initially at breakup speed and, after some 10^5 to 10^6 yrs, it has been spun down to 10% of it despite its contraction and mass accretion.

ReX-winds seem therefore to offer a serious possibility to brake down protostars (to my knowledge, there is no other calculated model in the literature). Note that ReX-winds are probably intermittent by nature because of the unavoidable radial drift of the reconnection site around the corotation radius (there is no ReX-wind whenever $r_X < r_{co}$). Dynamically speaking, such an unsteady "wind" should be better described as bullets flowing inside the hollow disc wind. Remarkably the basic features of X-type configurations remain if the stellar dipole is inclined: one would observe in that case precessing bullets channeled by the outer disc wind. Heavy numerical simulations will be required to test and further analyze this scenario.

References

1. Anderson, J. M., Li, Z. -Y., Krasnopolsky, R. & Blandford, R. D. 2005, ApJ, 630, 945
2. Balbus, S. A. 2003, ARA&A, 41, 555
3. Basu, S. & Mouschovias, T. C. 1994, ApJ, 432, 720
4. Basu, S. & Mouschovias, T. C. 1995, ApJ, 453, 271
5. Blandford, R. D. & Payne, D. G. 1982, MNRAS, 199, 883
6. Cabrit, S., Edwards, S., Strom, S. E. & Strom, K. M. 1990, ApJ, 354, 687
7. Cao, X. & Spruit, H. C. 2002, A&A, 385, 289
8. Casse, F. & Ferreira, J. 2000, A&A, 353, 1115
9. Casse, F. & Ferreira, J. 2000, A&A, 361, 1178
10. Casse, F. & Keppens, R. 2002, ApJ, 581, 988
11. Casse, F. & Keppens, R. 2004, ApJ, 601, 90
12. Contopoulos, J. 1994, ApJ, 432, 508
13. Contopoulos, J. & Lovelace, R. V. E. 1994, ApJ, 429, 139
14. Donati, J. -F., Paletou, F., Bouvier, J. & Ferreira, J. 2005, Nature, 438, 466
15. Fendt, C. & Elstner, D. 2000, A&A, 363, 208
16. Ferreira, J. 1997, A&A, 319, 340
17. Ferreira, J. 2002, in "Star Formation and the Physics of Young Stars", J. Bouvier and J. -P. Zahn (eds), EAS Publications Series, astro-ph/0311621, 3, 229
18. Ferreira, J. & Casse, F. 2004, ApJ, 601, L139
19. Ferreira, J., Dougados, C. & Cabrit, S. 2006, A&A, 453, 785
20. Ferreira, J. & Pelletier, G. 1993, A&A, 276, 625
21. Ferreira, J. & Pelletier, G. 1995, A&A, 295, 807
22. Ferreira, J., Pelletier, G. & Appl, S. 2000, MNRAS, 312, 387

23. Ferreira, J., Petrucci, P. -O., Henri, G., Saugé, L. & Pelletier, G. 2006, A&A, 447, 813
24. Frank, J., King, A. & Raine, D. J. 2002, Accretion Power in Astrophysics: Third Edition (Cambridge University Press)
25. Garcia, P. J. V., Cabrit, S., Ferreira, J. & Binette, L. 2001, A&A, 377, 609
26. Gounelle, M., Shu, F. H., Shang, H., et al. 2006, ApJ, 640, 1163
27. Hartigan, P., Edwards, S., & Ghandour, L. 1995, ApJ, 452, 736
28. Heiles, C., Goodman, A. A., McKee, C. F. & Zweibel, E. G. 1993, in Protostars and Planets III, ed. E. H. Levy & J. I. Lunine, 279–326
29. Heyvaerts, J. & Norman, C. 1989, ApJ, 347, 1055
30. Heyvaerts, J. & Norman, C. 2003, ApJ, 596, 1270
31. Königl, A. 2004, ApJ, 617, 1267
32. Königl, A. & Pudritz, R. E. 2000, in in "Protostars and Planets IV", Mannings, V., Boss, A. P., Russell, S. S (eds), Univ of Arizona Press, 759-+
33. Königl, A. & Wardle, M. 1996, MNRAS, 279, L61
34. Livio, M. 1997, in ASP Conf. Ser. 121: IAU Colloq. 163: Accretion Phenomena and Related Outflows, ed. D. T. Wickramasinghe, G. V. Bicknell, & L. Ferrario, 845
35. Long, M., Romanova, M. M. & Lovelace, R. V. E., 2005, ApJ, 634, 1214
36. Lubow, S. H., Papaloizou, J. C. B. & Pringle, J. E. 1994, MNRAS, 267, 235
37. Lubow, S. H., Papaloizou, J. C. B. & Pringle, J. E. 1994, MNRAS, 268, 1010
38. Matt, S. & Pudritz, R. E. 2005, MNRAS, 356, 167
39. Merloni, A., Heinz, S. & di Matteo, T. 2003, MNRAS, 345, 1057
40. Najita, J. R. & Shu, F. H. 1994, ApJ, 429, 808
41. Narayan, R., Mahadevan, R. & E., Q. 1998, in "The Theory of Black Hole Accretion Discs", eds. M. A. Abramowicz, G. Bjornsson, and J. E. Pringle astro-ph/9803141
42. Ogilvie, G. I. & Livio, M. 1998, ApJ, 499, 329
43. Okamoto, I. 2003, ApJ, 589, 671
44. Ostriker, E. C. 1997, ApJ, 486, 291
45. Ostriker, E. C. & Shu, F. H. 1995, ApJ, 447, 813
46. Ouyed, R., Clarke, D. A. & Pudritz, R. E. 2003, ApJ, 582, 292
47. Pelletier, G. & Pudritz, R. E. 1992, ApJ, 394, 117
48. Pesenti, N., Dougados, C., Cabrit, S., et al. 2004, A&A, 416, L9
49. Shakura, N. I. & Sunyaev, R. A. 1973, A&A, 24, 337
50. Shang, H., Glassgold, A. E., Shu, F. H. & Lizano, S. 2002, ApJ, 564, 853
51. Shang, H., Shu, F. H. & Glassgold, A. E. 1998, ApJ, 493, L91
52. Shu, F., Najita, J., Ostriker, E., et al. 1994, ApJ, 429, 781
53. Shu, F. H., Najita, J., Ostriker, E. C. & Shang, H. 1995, ApJ, 455, L155
54. Shu, F. H., Shang, H., Gounelle, M., Glassgold, A. E. & Lee, T. 2001, ApJ, 548, 1029
55. Vlahakis, N., Tsinganos, K., Sauty, C. & Trussoni, E. 2000, MNRAS, 318, 417
56. Wardle, M. & Königl, A. 1993, ApJ, 410, 218
57. Zanni, C., Ferrari, A., Massaglia, S., Bodo, G. & Rossi, P. 2004, Ap&SS, 293, 99

Stellar Wind Models

Christophe Sauty

Observatoire de Paris, L.U.Th., 92190 Meudon, France
christophe.sauty@obspm.fr

Abstract. I present here some ideas on how jets from low mass stars may evolve as the star evolves from class 0 to the main sequence through classes I, II, III. Analytical models and simulations suggest that the ejection start very early in the life of a low mass star from the edge of the disk. Then, the jet is progressively ejected from a more central part of the system composed by the star and its accretion disk. Once the disk itself evaporates, the jet becomes a mere wind from the star which has reached the main sequence. This wind should be similar to the well known solar wind. To illustrate this point, we show specific applications of meridionally self-similar models to jets from T Tauris with a low mass accretion rate, as well as for the solar wind. We also present numerical simulations of turbulent stellar jets surrounded by a magnetized disk wind, which clearly show that the stellar jet may be an essential ingredient in preventing too fast and too tight collimation of

C. Sauty: *Stellar Wind Models*, Lect. Notes Phys. **723**, 209–224 (2007)
DOI 10.1007/978-3-540-68035-2_9   © Springer-Verlag Berlin Heidelberg 2007

the disk jet. At this point, note that I shall call in the rest of the present review **"stellar jet"**, the component of the wind originating from the star and its vicinity, by opposition to the "disk wind" or "disk jet" that corresponds to the component emerging from the disk and more specifically from the Keplerian disk.

The analytical solution can be further extended to the relativistic domain. The same dichotomy seems to exist for extragalactic jets. While sub relativitic winds from Seyferts may correspond to radial winds, radio loud jets are comparable to YSO jets. Nevertheless the distinction between FRI and FRII jets may be more a problem of environment and efficiency of the magnetic rotator as FRI jets evolve in a rich external medium, while the opposite holds for FRII.

1 Introduction

1.1 Acceleration, Collimation and Source of the Jets

Basic of MHD wind theory is well explained in other lectures of this volume (see in particular Tsinganos for the notations). We shall here recall various results on stellar jet models.

The outflowing plasma is ruled by the usual three MHD equations, mass, momentum and energy conservations combined with Maxwell equations and Ohm's law for an ideal flow of infinite conductivity. It has been suggested [14] that these outflows may be formed simultaneously from the central star and the surrounding disk. Of course the jet originating from the star is thermally driven while the jet originating from a magnetized disk is more likely to be magneto-centrifugally driven. In both cases the eventual collimation of the wind into a jet may be of thermal origin because of the dense environment of the molecular cloud, but is more likely of to be of magnetic origin. A sketch is displayed in Fig. 1 to show the 3D structure of the outflowing plasma as well as a projection in the poloidal/meridional plane. It shows in particular that because of flux freezing, in steady, axisymmetric flows, streamlines and fieldlines are roped on the same flux tubes of constant mass and magnetic flux, respectively Ψ and A, such that in the poloidal or meridional plane, the poloidal components of the velocity and the magnetic fields are parallel. Besides the mass flux to magnetic flux ratio, $\Psi_A = \frac{d\Psi}{dA}$, other quantities are conserved along a given flux tube $A = $ cst. These integrals are the total angular momentum L, the corotation frequency Ω which is the angular velocity of the magnetic footpoints anchored into the source, and the total specific energy $\mathcal{E}$.

Various models have been put forward to explore the basics of MHD acceleration and collimation of winds into jets either analytically or through numerical simulations (see [4] and, in this volume, the lectures by Tsinganos, Ferreira and Lery). Analytically, disk winds have been studied for a long time in the past using the radially self-similar models extended from the original Blandford & Payne (1982) model. More recently, meridionally self similar models have been studied as they are the only way to study analytically the

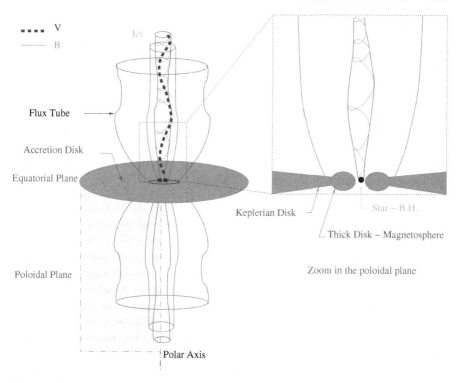

Fig. 1. Sketch of a magnetic flux tube emerging either from an accretion disk or the central corona of a star or a black hole. If the system is steady and axisymmetric, streamlines (V) and field lines (B) are roped on the same flux tubes. The plasma is accelerated and collimated further out

3D structure of axisymmetric winds emerging from the central star and the surrounding inner disk [14, 16].

At the present time, the question of the source of the jet remains among the unresolved problems though we have some hints. First, this source is probably not unique in the sense that the different regions of the disk and the stellar magnetosphere participate in launching the jet. Second, it probably evolves during the whole evolution of the central star. It seems clear though that the different parts of the accretion disk play an important role. In other words, at different stages, the external part of the disk, its inner radii or the central star itself are responsible for ejecting material. This evolution may be linked with the various classes of objects from class 0 to class III and the main sequence. This will be discussed later on. We shall first make a point on what analytical models from stellar jets can tell on the collimation itself before proceeding to the general evolution scenario.

1.2 Criterion for the Collimation of Jets

In order to treat almost analytically the partial differential equations of the MHD theory, we assume that there is a similarity in latitude such that if we know the evolution along one streamline, we can deduce at each latitude the evolution of all streamlines just by multiplying all quantities with a scaling factor depending only on the colatitude. Thus the system reduces to a system of ordinary equations with radius. The same is true for radially self similar solutions for disk wind where all quantities scale with radius instead of colatitude and the ordinary equations are a set of equations for the colatitude (see Tsinganos' lecture in this volume). Alternatively, the jet may also emerge from the connection between the disk and the star as suggested by Shu et al., 1994 for X-wind models (see also [17, 18], and references therein for more recent development of X-wind, in particular the calculation of the emission of such winds).

Considering the collimation, Heyvaerts & Norman (1989) have first established that for pure stellar outflows, the asymptotic shape of the jet could be linked with the current distribution inside the jet. Thus, winds with vanishing electric intensity and current density should become paraboloidal while current density carrying jets should be cylindrical. Finally, jets enclosing a non zero intensity but with null density current should become radial. These conclusions however do not hold for flows with strong discontinuity and current sheets or if they do not fill all space. In particular, the non polytropic meridionally self similar flows discussed hereafter all solutions are carrying a non zero current intensity despite the fact that they are either cylindrical or radial. There is no contradiction with the previous theorems because for instance, the solutions do not fill all space. In particular, cylindrical solutions are surrounded by a disk wind. Radial solutions do have either a null velocity on the last streamline connected to the star, which is in this case the asymptotically equatorial line, or an equatorial current sheet (an electric discontinuity disregarded in the previous analysis). Shang et al. (1998) also suggested that in X-wind the formation of a very dense axial wind could produce an apparent cylindrically collimated jet even though the real wind is only paraboloidal. It is however not completely obvious if this can be done on very large distances considering that X-wind models do not treat consistently the inner stellar wind.

In the case of meridionally self-similar models of cosmical MHD outflows, a rather general criterion for the collimation of winds into cylindrical jets has been established [15]. According to this criterion, if there is an excess of volumetric energy along a non polar streamline with respect to the axis, the outflow collimates asymptotically into cylinders. This is quantified by the parameter ε',

$$\varepsilon' = \frac{\rho(r, A)\tilde{E}(A) - \rho(r, \text{pole})\tilde{E}(\text{pole})}{\rho(r, A)L(A)\Omega(A)} , \tag{1}$$

which equals to the difference of converted energy $\tilde{E}$ of a line of magnetic flux A (i.e. the total energy once we have substracted the thermal content that remains at infinity) compared to the polar line and normalized to the energy of the magnetic rotator $L\Omega$. We recall (see above) that L is the angular momentum and Ω the corotation frequency of the streamline of magnetic flux A. We use here spherical coordinates $[r, \theta, \varphi]$. The parameter ε' is positive for collimated solutions and negative for non collimated winds. We notice that $\tilde{E}$ is the energy effectively used to accelerate and collimate the flow.

The parameter ε' has two contributions, one thermal μ and another magnetic ε, thus generalizing the usual criterion for fast vs. slow magnetic rotators, by taking into account thermal confinement as well,

$$\varepsilon' \equiv \mu + \varepsilon \,, \tag{2}$$

$$\mu \propto \frac{P(r, A) - P(r, \text{pole})}{P(r, \text{pole})} \,, \tag{3}$$

where $P(r, A)$ is the pressure along the line of magnetic flux A and $P(r, \text{pole})$ the pressure along the polar axis. Hence, μ measures the collimation due to the pressure gradient and in these models it is proportional to the relative variation of pressure across the outflow. When μ is positive (negative) the jet is underpressured (overpressured respectively) at the base of the flow. The magnetic parameter ε is

$$\varepsilon = \frac{L\Omega - E_{\text{R},o} + \Delta E_{\text{G}}^*}{L\Omega} \,, \tag{4}$$

where $E_{\text{R},o}$ is the rotational energy that tends to decollimate the wind because of the centrifugal force, and

$$\Delta E_{\text{G}}^* = -\frac{\mathcal{G}\mathcal{M}}{r_o} \left[1 - \frac{T_o(A)}{T_o(\text{pole})} \right] \,, \tag{5}$$

where $\mathcal{G}$ is the gravitational constant, $\mathcal{M}$ the mass of the star and T_o the temperature at the base of the flow r_o. The quantity ΔE_{G}^* in (5) corresponds to the gravitational potential well which is not compensated by thermal acceleration and thus must be supplied by the magnetic rotator in order to allow ejection. In other words, the parameter ε measures the quantity of energy of the magnetic rotator which is left once we have substracted the part that helps accelerating the flow. If there is an excess of such energy, the plasma is magnetically collimated. Thus, a fast magnetic rotator (large $L\Omega$) is not necessarily an **efficient magnetic rotator** ($\varepsilon > 0$) if magnetocentrifugal acceleration is important. Conversely a slow magnetic rotator (low $L\Omega$ as in T Tauri stars) can be either efficient ($\varepsilon > 0$) or inefficient ($\varepsilon < 0$) depending on the efficiency of thermal acceleration.

Efficient magnetic rotators are found if there is enough extended heating around the polar axis or if the magnetic field is strong such that the extraction of Poynting flux in the magnetocentrifugal acceleration is not too strong. In this case the wind collimate into a jet because of the toroidal magnetic pinching as in T Tauris. Conversely, in inefficient magnetic rotators, the winds either collimate thermally or remain conical, like the solar wind. This leads us to the following section where we shall discuss a general scenario for the evolution of the origin of the wind in young low mass stars (Fig. 2).

1.3 A Scenario for the Evolution of Jets

In the very early stage of class 0 objects, the jet may form before the star itself is born, i.e. as soon as the protostellar core exists. In class 0 (a), it has been suggested that the jet originates from the edge of the disk where there is still free falling material [5]. This is somewhat not very different from the idea of a recirculation around the proto stellar core suggested by Lery et al. (see this volume).

Then from class I to class II, the ejection proceeds towards the center of the disk, both radially self similar disk wind models and simulations seem to confort this idea (see Ferreira et al. and Tsinganos et al. this volume). Those models provide the right luminosity and mass loss rate (up to $10^{-7} M_\odot/yr$) provided that there is a small corona on top of the disk to launch the material which is further accelerated by the conversion of Poynting flux into kinetic energy as "a bead on a wire". In fact up to the Alfvén radius the magnetic

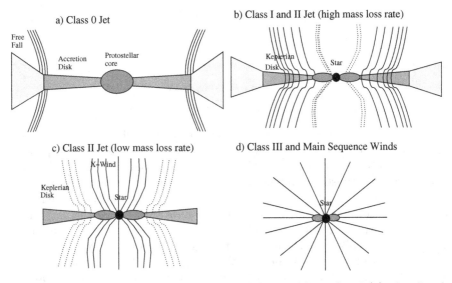

Fig. 2. Sketch of the evolution of the source of the wind from class 0 (**a**), class I and II with high ejection rates (**b**), class II with low mass loss rate (**c**) and main sequence. We see that the ejection proceed from the external part towards the central star as time goes

field forces the plasma to corotate with the Keplerian frequency Ω and because the accreted magnetic fieldlines are inclined, they act as inclined rigid wires on which the plasma is ejected pushed by the centrifugal force that exists in the corotating frame. This mechanism is also refered as magnetocentrifugal driving.

During the phase of class II objects, the jet may originate from the inner 3 stellar radii of the inner disk with a small contribution (10% of the star). This point will be illustrated more specifically in the rest of this review. At this stage, the magnetocentrifugal driving still exist but compete with the thermal driving. In fact the hot atmosphere around the star even if it is only a tenth of the solar coronal temperature can efficiently push the gas by creating a strong pressure gradient. Of course at this stage the mass loss rate is smaller, of the order of $10^{-9} M_\odot/yr$.

Eventually the disk evaporates and the star remains from class III (Weak T Tauri stars) to the main sequence with a non collimated wind similar to the famous solar wind with low mass loss rates, typically $10^{-14} M_\odot/yr$. This is one major advantage of meridionally self-similar models that they can also reproduce the structure of the quiet solar wind as well as they mimic the stellar jet core of T Tauri jets.

2 Stellar Jets and Winds

2.1 A Model for RY Tau and Low Ejection Mass Stars

As we mentioned, we will not discuss here the various models for disk winds as it is well exposed elsewhere in this volume. Let us just recall that disk wind models are well appropriated for early stage jets from class I and II. Most of the acceleration and the collimation of these jets are of magnetic origin by conversion of the Poynting flux. Nevertheless, the corona of the disk plays a crucial role as it is the coronal temperature and extension that produce enough mass loss rate and allow the solution to cross the all critical points appearing in MHD theory (e.g. [6]).

Sauty & Tsinganos 1994 have proposed, using meridionally self similar models, that for some T Tauris the inner part of the disk may play a crucial role combined with the stellar inner jet. This is some how very close to the X-wind models proposed by Shu et al. (see [18] and references there in) except for one essential ingredient. In X-winds, the geometry of the magnetic field at the connection between the stellar magnetosphere and the disk is a fan, see Figs. 1 and 4 in Shu et al. (1994). Here it is a real X point (see Figs. 8, 10 and 11 in Sauty & Tsinganos, 1994), such that the ejection from the disk is not concentrated in one single point but spread within the disk as in disk wind models. Figure 4 displays two sketches of the magnetic topologies of the two types of model. It clearly shows the differences between the X point of self-similar models and the fan of X-winds. Furthermore, the fan topology of the X-wind is suspected to be highly unstable by several authors (e.g. [2], see also Ferreira in this volume).

Using data from RY Tau we have constrained the parameters in the following way using data from Verdugo & Gomez de Castro (2001) for this object. For RY Tau we know the radius of the star (3 solar radii) and the rotation period of 24 days. We also know the asymptotic speed of the jet ($\approx$ 200 km/s) and UV lines indicates the existence of a shock at 38 AU which is spherical in shape. The density before and after the shock is known ($\approx 10^3 \text{cm}^{-3}$ and 10^4cm^{-3} respectively). Following the original paper, we assumed in the first place that the UV shock corresponds to the first recollimation point in the solution and we use it shape to estimate the location of maximum expansion of the solution. Note that meridionally self-similar solutions reproduce usually oscillating jets as explained in Vlahakis & Tsinganos (1998). Using those contrains the solution obtained is shown in Fig. 3a where we plot its three dimensional structure (with a logarithmic scale along the jet axis). The solution reproduced the main features except for the high temperature (1 million degrees!) in the asymptotic part. The high effective temperature may be explained if there is a lot of ram pressure or a magnetic pressure from Alfvén waves. However by lowering slightly the initial pressure, we were able to obtain a more satisfactory solution displayed in Fig. 3b. We have dropped the idea that we may reproduce the shock at the recollimation point as the shock is in any case not stationnary. Instead the jet that was previously both magnetically and thermally confined, still starts underpressured and thermally confined but becomes overpressured around 100 AU. Thus, the asymptotic collimation of this second solution has to be purely magnetic. Moreover the effective temperature in this solution is quite reasonable and can be considered as the kinetic temperature. It decreases continuously from 10^5 K at the stellar surface to 10^3 K asymptotically.

By rescaling the various parameters, we were able to reproduce, with the same numerical solution, various jets from T Tauris. As a matter of fact, all the solutions for T Tauri jets obtained seems to have a marginally efficient magnetic rotators $\varepsilon \approx 0$, which means that the toroidal magnetic field is collimating the jet but the force is weak such that the transverse extension of the jet (tens of AU) is very large compared to the radius of the underlying source.

The main output of these solutions is that the mass loss rate in the jet is in all cases around $10^{-9} M_\odot$ if the solution extends up to 3 stellar radii within the disk. The stellar jet itself is only $10^{-10} M_\odot$. Thus for T Tauri jets, the stellar component cannot explain by itself the observed jets and the dominant contribution seems always to come from the disk. However, in the case of low mass loss rates, typical of the late stages of T Tauris, only the very inner part of the disk contributes.

This leads to two conclusions. First the contribution of the disk is crucial even in such low accreting mass objects such as RY Tau. Moreover, we may have here a clue on the difference between jets from classical T Tauri Stars

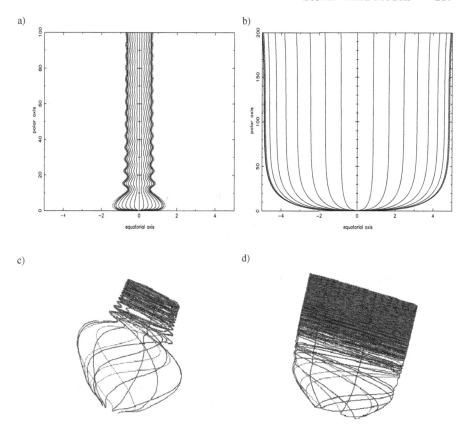

Fig. 3. We plot the morphology of two jet solutions for RY Tau. The solution plotted in the poloidal plane in (**a**) and in 3D for (**c**) shows recollimation at 38 AU. Parallely, the solution plotted projected in the poloidal plane in (**b**), and in 3D in (**d**). In (**b**) and (**d**), the new solution is obtained lowering slightly the initial pressure compared to (**a**) and (**c**). The recollimation point disappear. Instead the jet goes from underpressured to overpressured around 100 AU. The vertical axis of (**a**) and (**b**) is linear while it is logarithmic in (**c**) and (**d**), which explains the difference in the apparent morphologies. In (**a**) and (**b**) the units are in Alfvén radius

(noted CTTS hereafter) and weak T Tauri Stars (noted WTTS hereafter). In both cases we conjecture that the stellar jet is very similar. However as WTTS are not connected to the disk, the total mass loss rate remains of the order of $10^{-10} M_{\odot}$, which is almost not detectable. Conversely, jets from CTTS like RY Tau would have contributions from the inner disk which give a mass loss rate 10 times higher. It would not be surprising if indeed the inner stellar jets of CTTS and WTTS were similar, as we cannot see major differences between the stars themselves, in terms of rotation and magnetic activity. Conversely the contribution of the disk seems crucial to explain the difference in the total mass loss rate observed.

a)   b)

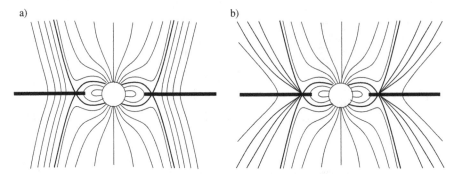

Fig. 4. Sketch in the poloidal plane of two magnetic topologies. In (**a**) the magnetic configuration in meridional self-similar winds which connect naturally to the configuration of disk wind models and in (**b**) the magnetic configuration of X-wind models. Lines in bold show the main differences between the two topologies

2.2 An Hybrid Model for the Solar Wind

The same model can be applied to Ulysses data of the solar wind at minimum [13]. The solution could not reproduce the high magnetic field measured at one AU, despite the correct values at the solar surface (roughly 1 Gauss) and the correct velocity (700 km/s) and density (a few particles per cm^{-3} obtained at one AU. The modelling has been improved [1] combining two models. The same model is used from the solar surface up to the Alfvén surface in order to reproduce the flaring streamlines of the fast solar wind as shown in Fig. 5a). Beyond the Alfvén surface, a helicoidal solution is used from the self similar model developed by Lima et al. (1991). Thus the geometry of the wind is purely radial in the poloidal plane as shown in Fig. 5b. This solution reproduces fairly well the dynamical and magnetic quantities both at the solar surface and at 1 AU. As other models (kinetic ones or simulations) we

a)   b)

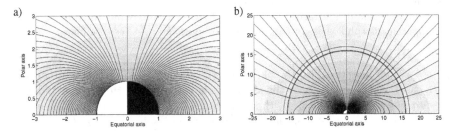

Fig. 5. Morphology in the poloidal plane of a model for the solar wind. Parameters are contrained from Ulysses data at minimum. The model reproduces the flaring of the lines up to the Alfvén radius. Beyond we use an helicoidal solution with radial poloidal streamlines. In (**a**), we plot a zoom of the inner streamlines showing the equatorial streamers, while in (**b**) we plot the overall structure at large distances. The grey scale corresponds to density contours

still have high temperatures for the protons. We may invoke ram pressure or Alfvén waves to explain the effective temperature that drives the wind. It can be shown that it is compatible with the data if the source of the acceleration is of magnetic origin (Alfvén waves) below the Alfvén surface and of turbulent origin beyond. At least the needed amplitude of the turbulent waves are of the order of the average magnitude of the magnetic and velocity fields.

Note that for the solar wind, the magnetic rotator is highly inefficient $\varepsilon \approx -50$, which is not surprising. It probably indicates that, as the star evolves and the magnetic field and the rotation decline, the magnetic efficiency in collimating is decreasing, at least in the last stages of the stellar formation. This goes also with a slowing down of the stellar magnetic braking.

3 Extensions of Stellar Jet Models

3.1 Towards Numerical Simulations

To go beyond analytical modelling, we may use numerical simulations. Tsinganos & Bogovalov (2001) have shown that a stellar wind can be collimated by the surrounding disk wind. There the origin of the acceleration of the stellar wind is not addressed as the stellar wind is a uniform supersonic outflow. Other numerical simulations have been performed using the VAC code (Meliani et al. 2006b). Previous simulations with this code did not include viscosity in the disk. The first step has been to include viscosity in the disk besides the magnetic diffusivity. It turns out, from the various simulations with different stellar mass loss rates, that the viscous transport of angular momentum in the disk remains negligible compared to the torque of the wind, even for a magnetic Prandtl number of one. This is so because the disk wind is very efficient in extracting angular momentum. The output obtained is not very different if a stellar jet is included in the middle with a mass loss rate only one percent of the total accreting mass (Fig. 6a) Here the simulation of Fig. 6, includes both turbulent viscosity in the disk and in the stellar jet. It is used as the source of the thermal driving of the inner stellar jet component. For higher mass loss rates from the star, the stellar jet starts opening significantly the disk wind to give a larger jet radius in better agreement with observations than the previous simulations. As we see in Fig. 6, this can be more than a factor of 2. As the lines bend more the magnetocentrifugal extraction from the disk becomes more efficient and the removal of angular momentum also increases.

These simulations also suggest that indeed the inner jet may be meridionaly self similar, while it has recently been shown that the disk wind can remain radially self similar [7].

3.2 Relativistic Models

Meridionally self-similar models can also be extended to study relativistic jets. Models for non rotating black holes, in a Schwarzschild metric have

a) 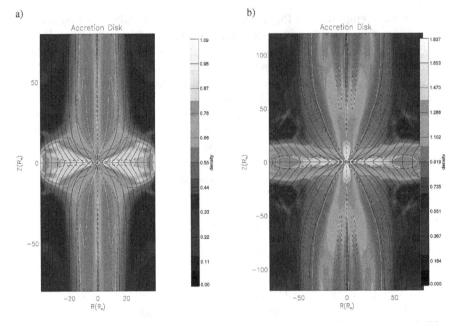 b)

Fig. 6. We plot the morphology in the poloidal plane of a stellar jet surrounded by an disk wind. In (**a**) the stellar mass loss rate is only 1% of the disk mass loss rate ($10^{-9} M_\odot/yr$ vs $10^{-7} M_\odot/yr$) while both are comparable in (**b**) ($10^{-7} M_\odot/yr$). The poloidal stream/fieldlines are in solid lines while the isocontours reflects the density distribution

been constructed (Meliani et al., 2006a). Such solutions can apply to relativistic winds rotating at subrelativistic rotation frequency, emerging from a hot corona around the central supermassive black hole (Fig. 1), which means to describe the inner spine jet component and not the overall jet which is better modelled by disk winds. However, this a very pedagogical exercice as it gives some insight into the complicated classification of AGN jets (Fig. 7, Urry & Padovani, 1995).

First of all, let us say that collimation criteria in terms of current distribution remain [9]. In self similar models, the criterion for collimation is also unchanged except for the introduction of the space curvature factor h of the Schwarzschild metric.

$$ds^2 = -h^2 c^2 dt^2 + \frac{1}{h^2} dr^2 + r^2 d^2\theta + r^2 \sin^2\theta d\varphi^2 , \qquad (6)$$

where

$$h = \sqrt{1 - \frac{2\mathcal{G}\mathcal{M}_\bullet}{c^2 r}} = \sqrt{1 - \frac{r_G}{r}} , \qquad (7)$$

is the redshift factor induced by gravity at a distance r from the central black hole of mass $\mathcal{M}_\bullet$, expressed in terms of the Schwarzschild radius

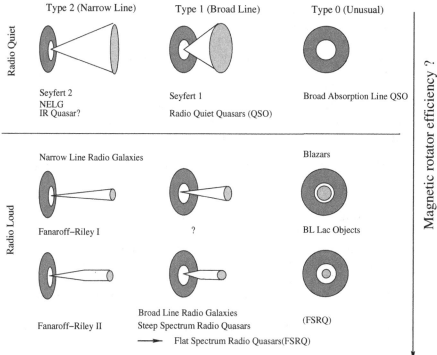

Fig. 7. Standard classification of AGN sources following Urry & Padovani, 1995. The horizontal axis can be explained as a change in the inclinaison of the viewing angle of the source with the line of sight. The vertical axis, as we suggest, may be linked to the efficiency of the underlying magnetic rotator to collimate the flow

$r_{\mathrm{G}} = 2\mathcal{G}\mathcal{M}_\bullet/c^2$. Note that the time line element or the lapse function is usually denoted by h_0 or α. In this review, we have simply used the symbol h.

Thus (4) giving the efficiency of the magnetic rotator is simply replaced by,

$$\varepsilon = \frac{\mathcal{E}_{\mathrm{R},o} + \mathcal{E}_{\mathrm{Poynt.},o} + \Delta\mathcal{E}_{\mathrm{G}}^*}{\mathcal{E}_{\mathrm{MR}}} \ , \qquad (8)$$

where $\mathcal{E}_{\mathrm{MR}} = h^2 L\Omega$ is again the energy of the magnetic rotator, $\mathcal{E}_{\mathrm{Poynt.}} = -h\varpi\Omega B_\varphi/\Psi_A$ is the Poynting flux, B_φ the toroidal magnetic field, Ψ_A the mass to magnetic flux ratio, using cylindrical coordinates $[\varpi, \varphi, z]$, and

$$\mathcal{E}_{\mathrm{R},o} = \frac{\mathcal{E}}{c^2} \frac{V_{\varphi,o}^2}{2} \ , \qquad (9)$$

is the rotational energy per particle. It is proportional to the specific rotational energy $V_{\varphi,o}^2/2$, with the factor $\mathcal{E}/c^2$ having the dimensions of a mass. We recall that $\mathcal{E}$ is the total specific energy of the plasma. Finally, $\Delta\mathcal{E}_{\mathrm{G}}^*$ is a term similar to the nonrelativistic case where it measures the excess or the deficit on a non polar streamline, compared to the polar one, of the gravitational energy per unit mass which is not compensated by the thermal driving [12, 15]. As in the classical case, ε measures the efficiency of the magnetic rotator to collimate the flow. Thus if $\varepsilon > 0$ we have an Efficient Magnetic Rotator (EMR) where magnetic collimation may dominate, while if $\varepsilon < 0$ we have an Inefficient Magnetic Rotator (IMR) where collimation cannot be but of thermal origin.

In fact, as already seen in numerical simulations of stellar relativistic jets [20], magnetic collimation of relativistic jets is harder to achieve because in the observer frame, the ultrarelativistic jet appears to be heavier. Conversely, relativistic jets are more efficiently thermally accelerated, because of space curvature and because the relativistic temperatures are more efficient (Meliani et al., 2004).

As displayed in Fig. 8, the solutions obtained are very similar to those for T Tauri jets. We may compare Seyfert winds to the solar wind. There, the efficiency of the magnetic rotator is so low that the wind remains uncollimated and in fact it seems also subrelativistic with typical velocities around 30,000 km/s. Conversely, Fanaroff-Riley I jets (FRI) from radio-loud galaxies are ultra-relativistic on the parsec scale but probably decelerate on larger scales. This last effect may be due to the rich host galaxies where the jet is embedded. Thus, the recollimated solutions correspond to decelerated ones with a peak of the Lorentz factor around 10 and thermal confinement. Conversely

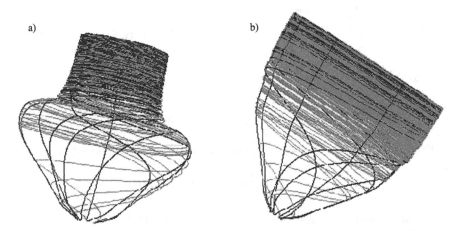

a)   b)

Fig. 8. We plot two relativistic solutions. In (**a**) the solution corresponds to the spine jet of a FRI, with a rich external medium. The jet is partially thermally confined and at the recollimation point the flow slows down. This is similar to real FRI jets where usually the kiloparsec scale jet is subrelativistic while the parsec scale jet is relativistic. In (**b**) the non recollimating solution with a continous acceleration up to relativistic asymptotic speed (with again Lorentz factors around 10) may correspond to ultrarelativistic jets of FRII where the environment is poor and the collimation has to be of magnetic origin

the collimated solutions without recollimation, correspond to a continuous acceleration with similar Lorentz factors and purely magnetic confinement, as the jet becomes overpressured. This situation is closer to the situation of FRII jets which are very powerfull, relativistic on all scales, and evolving in a poor external medium.

Thus, there may be a strong analogy between YSO jets and AGN jets though with a huge difference. The classification of AGN jets seems more related to difference in the environment and the magnetization of the rotating source, rather than due to a natural evolution from jets to winds as in young stars.

4 Conclusions

Thus, as a conclusion, it could be worth to combine radially self similar models for the disk wind and meridionaly self similar ones for the stellar jet to study analytically the interaction between the two component. The advantage of such a construction is that for the moment being numerical simulations remains at a high cost to be performed.

We have seen here that there are strong theoretical arguments to believe that the ejection proceeds towards the center of the system disk+ star. There are also observational evidences for that and this is discussed in other sections of this book.

Last but not least, studying jets from young stars and winds from main sequence stars can also help understanding jets on larger scales, as AGN jets. The analogy is worth studying both analytically and numerically.

Acknowledgement

I am very grateful to all my colleagues for their contribution to various part of the work presented in this review, in alphabetic order A. Aibeo, F. Casse, I. Contopoulos, J. Lima, Z. Meliani, E. Trussoni, K. Tsinganos and N. Vlahakis. I also thank the organisers for inviting me at this very enjoyable and fruitful JETSET school.

References

1. Aibeo, A., Lima, J.J.G., Sauty, C., 2006, A&A, accepted
2. Bardou, A., Heyvaerts, J., 1996, A&A, 307, 1009
3. Blandford, R.D., Payne, D.G., 1982, MNRAS, 199, 883
4. Bogovalov, S., Tsinganos, K., 2001, MNRAS, 325, 249
5. Contopoulos, I., Sauty, C., 2001, A&A, 365, 165
6. Ferreira, J., Casse, F., 2004, ApJ, 601L, 139

7. Gracia, J., Vlahakis, N., Tsinganos, K., 2006, MNRAS, 367, 201
8. Heyvaerts, J., & Norman, C., 1989, ApJ, 347, 1055
9. Heyvaerts, J., & Norman, C., 2003, ApJ, 596, 1240
10. Lima, J.J.G., Priest, E.R., Tsinganos, K., 2001a, A&A 371, 240
11. Meliani, Z., Sauty, C., Vlahakis, N., Tsinganos, K., Trussoni, E., 2006, A&A, 447, 797
12. Meliani, Z., Casse, F., Sauty, C., 2006, A&A, (submitted)
13. Sauty, C., Lima, J.J.G., Iro, N., Tsinganos, K., 2005, A&A, 432, 687
14. Sauty, C., Tsinganos, K., 1994, A&A, 287, 893
15. Sauty, C., Tsinganos, K., & Trussoni, E., 1999, A&A, 348, 327
16. Sauty, C., Trussoni, E., Tsinganos, K., 2002, A&A, 348, 327
17. Shang, H., Shu, F.H., Glassgold, A., 1998, ApJ, 493, 91
18. Shang, H., Lizano, S., Glassgold, A., Shu, F.H., 2004, ApJ, 612, 69
19. Shu, F.H., Najita, J., Ostriker, E., Wilkin, F., Ruden, S., Lizano, S., 1994, ApJ, 429, 781
20. Tsinganos, K., Bogovalov, S., 2002, MNRAS, 337, 553
21. Vlahakis, N., Tsinganos, K., 1998, MNRAS, 298, 777

Index